T & T Clark
Religion and the University Series

Series editors

Volume 6:
Our Common Cosmos: Exploring the Future of Theology, Human Culture and Space Sciences

Religion and the University Series

The Religion and the University series aims to facilitate a creative and imaginative role for the Christian theological perspective within the university setting, working from the premise that religious culture can make a valuable contribution to wider university education. Contributions are welcome and prospective editors and authors can gain further information at http://www.bloomsbury.com/uk/series/religion-and-the-university/

OUR COMMON COSMOS

Exploring the Future of Theology, Human Culture and Space Sciences

Edited by
Zoë Lehmann Imfeld and Andreas Losch

t&tclark
LONDON • NEW YORK • OXFORD • NEW DELHI • SYDNEY

T&T CLARK
Bloomsbury Publishing Plc
50 Bedford Square, London, WC1B 3DP, UK
1385 Broadway, New York, NY 10018, USA

BLOOMSBURY, T&T CLARK and the T&T Clark logo are trademarks of
Bloomsbury Publishing Plc

First published in Great Britain 2019

A catalogue record for this book is available from the British Library.

A catalog record for this book is available from the Library of Congress.

ISBN: HB: 978-0-5676-8016-7
ePDF: 978-0-5676-8017-4
ePUB: 978-0-5676-8018-1

Typeset by Deanta Global Publishing Services, Chennai, India
Printed and bound in Great Britain

To find out more about our authors and books visit www.bloomsbury.com
and sign up for our newsletters.

CONTENTS

LIST OF CONTRIBUTORS

Dr Conor Cunningham is Associate Professor in Theology and Philosophy, and Director of the Centre of Theology and Philosophy, University of Nottingham. He is the author of *Genealogy of Nihilism: Philosophies of Nothing and the Difference of Theology* (Routledge 2002), which has since been translated into Chinese. He is also the author of the award-winning book *Darwin's Pious Idea* (Wm. B. Eerdmans 2010). This too has been translated into a number of languages – thus far, Korean, Spanish and Russian. The third part of this trilogy is forthcoming: *Soul and the Marriage of Discourse: The Return of Scientia* (Wm. B. Eerdmans). Research for this book, of which his chapter in this volume is representative, began during his time as a fellow at CTI, Princeton. In addition, Cunningham was the writer and presenter of the multi-award-winning BBC documentary – *Did Darwin Kill God?*

Prof. Dirk Evers is Professor of Systematic Theology/Dogmatics at Martin-Luther-University of Halle-Wittenberg, Germany (since 2010). Professor Evers received his PhD in theology from Tübingen University in 1999, followed by his *Habilitation* in 2005. He has been President of the European Society for the Study of Science and Theology (ESSSAT) since 2014; he is editor of the journal *Philosophy, Theology and the Sciences*, and a member of the International Society for Science & Religion (ISSR). Professor Evers is also an ordained minister of the German Lutheran Church.

Dr Zoë Lehmann Imfeld was recently Mileva Maric Fellow at the Center for Space and Habitability (exact sciences), University of Bern. She is a senior lecturer in modern English literature at the University of Zurich and lecturer at the University of Bern. She is currently working on a postdoctoral project on the hermeneutics of space research in science-fiction narratives. She completed her PhD thesis in 2015, which was published in 2016 as *The Victorian Ghost Story and Theology: From Le Fanu to James* by Palgrave MacMillan (Springer-Verlag). Her thesis was awarded the SNF Marie Heim-Vögtlin Prize in 2016. In 2017 Zoë was a visiting fellow at the Center of Theological Enquiry, Princeton, as part of the NASA-funded project 'Inquiry on the Societal Implications of Astrobiology'. Zoë is co-editor with Peter Hampson and Alison Milbank of the volume *Theology and Literature after Post-Modernity*, also part of this Bloomsbury series.

Dr Andreas Losch is an award-winning theologian, specializing in the dialogue between the sciences, philosophy and theology. He is currently principal investigator of the project 'Ethics of Planetary Sustainability' at the University of Bern, Switzerland (www.planetarysustainability.unibe.ch). Losch is a member of the Center of Theological Inquiry, Princeton, New Jersey, and he serves in the councils of the European Society for the Study of Science and Theology and in the Karl Heim Society. He is also editor-in-chief of a German forum for dialogue between the sciences and theology. Andreas has most recently edited *What Is Life? On Earth and Beyond*, Cambridge University Press, 2017.

Prof. Tom McLeish, FRS, is Professor of Natural Philosophy at York University, where he works with chemists, engineers and biologists in universities and industries to connect material properties with their molecular structure. He also works on connections between science and policy, history (e.g. interdisciplinary re-examinations of medieval scientific treatises) and the humanities, resulting in his recent books *Faith and Wisdom in Science* (OUP 2014) and *Let There Be Science* (LionHudson 2017). He was Pro-Vice-Chancellor for Research at Durham University (2008–14) and is Chair of the Royal Society's Education Committee. He has been a reader (lay preacher) in the Anglican Church since 1993.

Prof. Markus Mühling is Professor of Systematic Theology at the Protestant University Wuppertal/Bethel, Germany. He has been a member of the CTI, Princeton, and has held guest professorships in Aberdeen (2009/10) and Oxford (2017). Professor Mühling's monographs include *Einstein und die Religion* (Vandenhoeck 2011), *Liebesgeschichte Gott* (Vandenhoeck 2013), *Resonances: Neurobiology, Evolution and Theology* (Vandenhoeck 2014) and *The T&T Clark Handbook of Christian Eschatology* (2015).

Dr Erik Persson has a PhD in Practical Philosophy from Lund University (2009). His main research interest is applied ethics, especially environmental ethics and ethical issues in relation to science and emerging technologies. Most recently he has been involved in two projects dealing with different aspects of life: as principal investigator for the project 'A Plurality of Lives', funded and hosted by the Pufendorf Institute for Advanced Studies at Lund University, and as Research Fellow in the project 'Inquiry on the Societal Implications of Astrobiology', hosted by the Center of Theological Inquiry, Princeton, and funded by NASA. His chapter in this book emerges from the latter project.

Dr Carl B. Pilcher served as the director of the NASA Astrobiology Institute (NAI), headquartered at the NASA Ames Research Center in Mountain View, California, from 2006 until his retirement in January 2013. He served again as part-time interim director from August 2014 to May 2016. Prior to his terms as NAI Director, he served for eighteen years at NASA Headquarters in Washington, DC, in numerous capacities, including Senior Scientist for Astrobiology and Science Director for Solar System Exploration. He began his career at the Institute for Astronomy of the University of Hawaii, studying planets and moons in the outer solar system. He has BS and PhD degrees in chemistry from the Polytechnic Institute of Brooklyn and MIT, respectively, and a Master of Public Affairs degree from the Woodrow Wilson School of Princeton University. In 2013 he was awarded the NASA Distinguished Service Medal, the highest honour bestowed by NASA on a federal employee.

Dr Fabien Revol is a member of the Interdisciplinary Centre for Ethics, holder of the Jean Bastaire Chair on Integral Ecology and the assistant coordinator of the Chair on Science and Religion at the Catholic University of Lyon. He teaches eco-theology, science and religion, and ethics in the different faculties of his university. He is the author of several books on science and religion dialogue especially the topic of continuous creation: *Le temps de la création*, Paris, Cerf, 2015, *Le concept de création continuée dans l'histoire de la pensée occidentale*, Paris, Vrin, 2017, and on eco-theology, such as *Pour une écologie de l'espérance*, Lyon, Peuple Libre, 2015, and *Avec Laudato si' devenir acteur d'écologie intégrale* (ed.), Lyon Peuple libre, 2017.

Prof. Knut-Willy Sæther is Professor of Religious Studies at Volda Department of Religious Studies, at Volda University College Norway. His main interest of research is interdisciplinary, involving philosophy, aesthetics, theology and science. Sæther earned his doctoral degree from the Norwegian University of Technology and Science (2005) on the dialogue between theology and science. He served as Scientific Programme Officer at European Society for the Study of Science and Theology (ESSSAT), 2010-2018. Among his publications are *Traces of God* (2011), *Skjønnhet og tilbedelse* (*Beauty and Worship*) (2013), *Kristen Spiritualitet* (*Christian Spirituality*) (2013) and *Naturens skjønnhet* (*Beauty of Nature*) (2017).

Dr Howard A. Smith is a senior astrophysicist at the Harvard-Smithsonian Center for Astrophysics and a lecturer in the Harvard University Astronomy Department, and has been recognized by Harvard for excellence in teaching. He is an author of over 350 scientific publications. He is the author of the book, *Let There Be Light: Modern Cosmology and Kabbalah, a New Conversation between Science and Religion* (New World Library), and he writes and lectures on astrophysics and Judaism. Prior to coming to Harvard he was the chair of Astronomy at the Smithsonian's National Air and Space Museum; he also served as a visiting discipline scientist in Astrophysics at NASA Headquarters.

Prof. Robert (Bob) S. White FRS is Professor of Geophysics at Cambridge University (since 1989) and Director of The Faraday Institute for Science and Religion. He is a fellow of St Edmund's College, Cambridge, The Royal Society (1994) and the American Geophysical Union. In 2018 he was awarded a Gold Medal of the Royal Astronomical Society, in recognition of a lifetime's achievement in research. He leads a research group investigating the Earth's dynamic crust: in particular the way in which enormous volumes of volcanic rock are produced when continents and oceans rift apart, and the movement of molten rock under active volcanoes.

FOREWORD

Carl Pilcher
Blue Marble Space Institute

Like many kids of my post-Second World War generation who grew up to become scientists, I found science fiction a pathway to scientific thinking and imagination. As I reflect on the science fiction that left its greatest imprint on me, I discern a common theme of examining the human condition. I am haunted to this day by the image of Bob Shaw's Mr Hagan in *Light of Other Days*, sitting on a low wall in front of his stone farmhouse, watching his wife and child through windows of 'slow glass' years after their deaths in an auto accident. And by The Mule, the galactic conqueror in Isaac Asimov's Foundation Trilogy, who could reach into others' subconscious and 'adjust' their emotions as if turning a dial. And by Walter Miller's novice Brother Francis venerating and protecting with his life an ancient, mysterious, holy relic that inscribes the intention of his order's founder to pick up a 'pound pastrami, can kraut, six bagels' on his way home to wife Emma in *A Canticle for Liebowitz*.

And so perhaps it is not strange that through a circuitous path that included chemistry and planetary astronomy, international relations and government policy, I found myself working in astrobiology, the quest to understand the potential of the universe to harbour life beyond Earth. For this quest engages human awe and aspirations as well as sciences spanning the disciplinary spectrum to address age-old questions. *How did life begin? How did the complexity of life on Earth arise? Are we alone in the universe, or are there planets elsewhere with life? What is the future of life on Earth and beyond?*

Before the advent of modern science these questions (or more rudimentary precursors) were the domain of philosophers and theologians. The Greek atomists had a remarkably modern, pluralistic perspective, arguing that an infinite number of atoms led to 'infinite worlds both like and unlike this world of ours … . We must believe that in all worlds there are living creatures and plants and other things we see in this world' (Epicurus, 'Letter to Herodotus'). However, it was not this perspective but the hierarchical views of Aristotle that dominated thought for many centuries following. Aristotle argued that Earth was the centre of the cosmos and hence unique, a view that suited Christian scholars such as thirteenth-century Thomas Aquinas, who wrote 'When it is said … that many worlds are better than one, … this sort of better … does not belong to the intention of God … because for the same reason it could be said that if he made two, it would be better that there were three; and thus ad infinitum' (Aquinas, *Summa Theologiae*, I 47.3 ad 2).

The twentieth century saw the tools of modern science used to transform these questions from subjects of epistemological debate to objects of empirical investigation. The modern synthesis merged Darwin's insights into evolution with Mendel's into heredity to produce a new genetic understanding of evolution. Discoveries into the nature of atoms combined with new astronomical spectrometers led to an understanding of the nature of stars. The space age produced unprecedented views of other planetary bodies. And high precision astronomical techniques finally revealed the presence of long suspected planets around other stars. By the late twentieth century the stage was set for astrobiology. So when, on 7 August 1996, in the auditorium at NASA Headquarters in Washington, DC, several scientists announced possible evidence of ancient life in a meteorite from Mars, they sparked a coalescence of scientists, politicians and government officials that led to a new NASA's Origins Program, including founding the NASA Astrobiology Institute (NAI).

The NAI, which I was privileged to direct for most of a decade (2006–16), was charged with providing leadership to NASA spaceflight missions, educating the next generation of astrobiologists, sharing the excitement and discoveries of astrobiology with the public and developing the concept of a virtual institute. But first and foremost, the NAI was formed to bring together scientists from a wide range of disciplines to address astrobiology's fundamental questions. These disciplines span the biological, geological and astronomical sciences and include cross-cutting areas of physics and chemistry. But it was also recognized that bringing together scientists, however central, was not enough. Humanists are also vital participants in an endeavour addressing questions that resonate so deeply with so many, and which engages pressing issues facing society.

Perhaps the most important category of pressing issues engaged by astrobiology is ecological, including the interrelated challenges of climate change, sustainability and biodiversity. Human impact on the environment is so great that it has been suggested that we have entered a new geological era termed the *Anthropocene*, defined by the lasting imprint humans are making on the planet.[1] Another issue, raised by the era of space exploration, is planetary protection. With both governments and (soon) private companies launching spacecraft to other solar system bodies, how do we address potential contamination of extraterrestrial environments and threats, however unlikely, to life on Earth from returned materials? And of course the field of astrobiology, which seeks a scientific answer to the question *How did life arise on Earth?*, engages deeply held religious beliefs.

These issues and others are addressed in this book, which seeks to strengthen a nascent bridge between the science of astrobiology and the humanities. It is fitting that one of the editors (ZLI) is a literary scholar and theologian with a particular

1. See, for example, D. Grinspoon, *Earth in Human Hands: Shaping Our Planet's Future* (New York: Grand Central Publishing, 2016). Also A. Frank and W. Sullivan, 'Sustainability and the Astrobiological Perspective: Framing Human Futures in a Planetary Context', *Anthropocene* 5 (March 2014): 32–41.

interest in science fiction and the other a theologian (AL) who works at the interface of religion, ethics and science. For it is only by bringing these diverse perspectives and their practitioners into dialogue with one another, the scientific community and the public at large that we can realize the pluralism that is a prerequisite for the understanding sought by astrobiology.

INTRODUCTION

Zoë Lehmann Imfeld and Andreas Losch
University of Bern

We realize that the subtitle of this book may seem presumptuous. It was, and still is, daring to reach out from theological ground to the domain of the natural sciences, and even more so to move vice versa. Yet this is part of the story of how this book was conceived: in discussions with scientists, among theologians and philosophers, at the *Center of Theological Inquiry* (CTI) in Princeton, New Jersey, an independent think tank and reservoir for fresh thinking. It was here that both editors took part in the *Inquiry on the Societal Implications of Astrobiology*. Co-funded by NASA and the Templeton Foundation, this inquiry continues a history in which NASA has reached out to and engaged with representatives from the humanities in general and religious leaders in particular. As NASA's Carl B. Pilcher, an enthusiastic scientist with an open mind, told us: 'Many NASA folks are interested in the societal implications of the work they do. They recognize—indeed many went to work for NASA because—their work addresses aspirations of individuals and society as a whole.' Mary Voytek in particular must be mentioned, Senior Scientist for Astrobiology at NASA, for maintaining this relationship, an effort which in itself has an element of daring. While NASA has navigated a steady course through various storms, the environment in which it operates has shifted so radically over the past couple of decades that what is normal for NASA is now seen as counter to prevailing winds. In the current political climate (and especially the current US administration) an increasing number of natural scientists perceive that 'carrying on as usual' with earth and space sciences research will not suffice. Their mission demands special action, a response to a political climate that appears to privilege short-term financial returns over this planet's sustainability.

Society needs science to understand what this world is about. And the natural sciences need to deal with the variety of religious beliefs in this regard, as we look to our future interaction with this planet and its neighbours.[1] The task here, however, is not simply to insist on cooperation, but to investigate the forms which such cooperation might take, navigating the tension between the (increasingly urgent)

1. Andreas Losch (2018) also explores the ethical future of planetary sustainability in 'The Need of an Ethics of Planetary Sustainability'.

need for hospitable interaction and the need to maintain the identity and legitimacy of distinct disciplines. The complexity of such a task is reflected in the broad, and often provocative, scope of responses to it contained within this volume.

Such an exchange demands that we address challenges head-on. Hospitality to other perspectives must be bolstered by intellectual rigour. For instance, the International Society for Science and Religion (ISSR), modelled after the Royal Society, issued a statement against intelligent design, calling it 'bad science and bad theology' as well. This sounds compelling at first, but here, the ISSR is wrong. Intelligent design is certainly bad theology, but as any biologist will tell us, it is not science *at all*. Theologians (and as it seems, religious scientists alike) often see no difficulty with a concept in which the divine intervenes in evolutionary affairs. Yet the history of science teaches us that this is not a sufficient description of natural phenomena, certainly not within the sciences. As we see in the pages of this volume, as theologians, we are free to assume that God is creative somehow, but we must also respond to a world which 'makes itself'.[2]

The function of theology within the context of science is the focus of Part I of this book. The volume starts by acknowledging the plurality of views on the contemporary interaction between science, ethics and theology, and shows how a constructive dialogue is needed within the current scientific climate.

As the following chapters demonstrate, however, scientific developments raise as many questions as they answer, many of these ethical and philosophical. The natural sciences require a 'language' with which to approach these questions. Part II offers a variety of languages for this purpose, from philosophy of science to literature, to aesthetics.

Science, however, is ultimately a human endeavour. Theology can, we argue, play a part in *scientific imagination*, providing good (or bad) ideas drawn from its mythologies that can then be put to the test of rigorous scientific method. Belief, however (be it religious or science-based), can also bias us, becoming a symptom of wishful thinking, or causing the 'adjustment' of historical examples in a way we would have liked it to be. Science as a sceptical enterprise (see also Losch 2017, 308), can act here as a purifying fire, which burns on the fuel of imagination.

'Our common future' was the title of the groundbreaking Brundtland report on sustainable development of our planet, and the theological questions inherent to the discussion were recognized by Pope Francis in his *Laudato Si*, with its subtitle of 'On care for our common home'. This volume extends these discussions to accommodate a dynamic and rapidly expanding set of disciplines. We ask: What will it take to shape the development of our efforts not only on Earth, but also in space in the same way, taking theology into account?

2. Charles Kingsley wrote to Darwin that 'he has gradually learnt to see that it is just as noble a conception of the Deity to believe that He created a few original forms capable of self-development into other and needful forms, as to believe that He required a fresh act of creation to supply the voids caused by the action of His laws' (Darwin 1860, 481).

This volume employs theology in a constructive approach to earth and space sciences. Here, we look to theology to be responsive to the rapid and often radical developments happening in this field. Related in their hospitality to a scientific imagination, these chapters represent and exemplify the tangled web that interdisciplinary endeavours weave. The role of religion outside of its own 'concerns' is a continuing debate, in wider society in general and scientific endeavour in particular.[3] Such interactions, however, must be hospitable on both sides. This volume seeks not to demonstrate that science 'needs' theology to function, nor that theology and the philosophical voice of the humanities 'need' to submit to the discourse of science to survive. Rather, the chapters collected here seek to explore the ways in which such conversations can be navigated, what kind of spaces can be created in which genuine exchange can take place.

Our Common Cosmos has been constructed so as to represent two distinct aspects of interdisciplinary exchange. Part I we have called 'Approaches', and in it theologians from a variety of backgrounds task themselves with finding a role for theological thinking in the natural and exact sciences. The juxtaposition of Markus Mühling's chapter 'Conversations along the Way: How and Why Science and Theology Need to Interact' and Dirk Ever's '"Good Fences Make Good Neighbours": Why the differences of science, religion and theology must not be blurred' is a testament to the intricacies of nurturing hospitable exchange while remaining alert to the integrity of each discipline and its own values. While Mühling describes a participatory construction of an interdisciplinary perspective, Evers finds the metaphor of perspective itself problematic, insisting not only on different perspectives, but on different topographies. Andreas Losch, meanwhile, revisits Robert John Russell's idea of creative mutual interaction, based on Barbour's graphical representations of exchange between science and theology, to redirect our attention to the channels of communication *between* disciplines, offering a democratized interpretation. Finally, Conor Cunningham calls for a radical reassessment of the hierarchy of the sciences, and sees the theo-philosophical tradition of *scientia* as a means to undermine what he sees as the imperialism of scientific reductionism. In each of these chapters, theology is offered as a discipline alert and responsive to the challenges posed to it by scientific methods and methodologies.

In Part II, which we have called 'Interactions', representatives from the humanities and natural sciences explore the ways in which their own disciplines can act as mediators, interpreters or perhaps necessary antagonists to the communication between science and theology. Such a grouping aims to highlight and encourage the spirit of *participation* with which this volume was envisaged. Bob White brings his perspective from geophysics in 'Sustainability: Interaction between Science, Ethics and Theology', with a timely reminder of the urgency and materiality of the themes discussed here. White depicts the vast scales with which

3. Rowan Williams 2012, for instance, examines the place for the religious voice in *Faith in the Public Square*.

his field is concerned as something also immanently human in scale. Fabien Revol uses the language and concepts of theology to describe biodiversity as an ethical imperative for humans embedded in their environment. For Knut-Willy Sæther, the language of aesthetics reconstructs the contested territory of the natural world in such a way as to afford both scientific and spiritual wonder. Zoë Lehmann Imfeld's chapter 'Imagination as Co-Creation: Science and Theology through the Lens of Science Fiction Literature' offers fiction as a means to escape the limits of our currently observable universe in order to give shape to the possibilities (both anticipated and feared) of our accelerating relationship with space.

Already, then, we see that the landscape between theology and science, so carefully plotted and mapped in section one, is being cultivated and tended. Erik Persson and Howard Smith then ask how we are to inhabit this landscape, Persson providing a guidebook from philosophy of science that describes various paths through encounters with astrobiology. Smith then uses astrophysics to confront us with an encounter of another kind, our seeming solitude in the cosmic landscape. To end our volume, Tom McLeish draws these narratives together to suggest a journey through this new territory that is both adventurous and contemplative.

In academia, 'interdisciplinary' scholarship continues to be seen as some sort of Holy Grail, although the quest can seem unobtainable. In particular the work of empirical scientists and theologians can often seem untranslatable to each other. The truth claims and premises of one scholarly field are often treated as a confrontation to the other – challenges to be overcome. And the chapters included here are challenging. The very form of 'interdisciplinarity' is often contested within these pages, each contributor attempting to map out a space in which the humanities and the natural sciences can communicate. Let's explore some examples of such a conscious conversation that happen within this book. Dirk Evers argues that only by first mapping out boundaries between disciplines can those disciplines then successfully interact. Each discipline demands its own 'topology', claims Evers, from which a comprehensive map of reality can be drawn. Indeed, for several of our contributors, the clear identification of individual disciplines is necessary if we are to compile any interdisciplinary map at all. Knut-Willy Sæther, as his title 'Aesthetics at the Intersection of Science and Theology' promises, invites the language and perspective of aesthetics to act as a mediator between disciplines. For Sæther, aesthetics provides a topology for the contested space of the natural world as both empirical and experiential. For others, the contested nature of this space becomes territorial, a space in which one discipline makes claims of dominance over another. In something of a challenge to the physics-oriented focus of this book, Conor Cunningham, in 'Who's Afraid of Reductionism's Wolf? The Return of Scientia' describes physics itself as an 'imperialistic' methodology, and makes a case for chemistry as a far more suitable candidate to bridge the gap between the humanities and the sciences.

Such challenges are a crucial part of the purpose of this volume. The instinct of the academy in reaching for interdisciplinary scholarship is often to try to collapse the distinctions between disciplines in order to do interdisciplinary work. The temptation is to keep discussions within a safe space, sticking only to common ground. In this way, the thinking seems to go, the disciplines themselves are not contested. Even more tempting, of course, is to insist on the autonomy of disciplines. Within such compartmentalization, disciplines not only have their own topology, as Dirk Evers describes here, but are, as Ian Barbour calls them, autonomous 'domains' (Barbour 2000, 17; see also Gould 2002). In this way each discipline keeps itself safe from the challenges of differing methodologies and differing 'truth-claims'. And yet this volume demonstrates that even domains can be challenged, and moreover can challenge the relationship between disciplines. Conor Cunningham uses the concept of domains from quantum field theory to undermine the hierarchical status of the exact sciences, for instance, and Andreas Losch asks us to resituate our attention from the domains themselves to the currents and channels running between them, in his chapter 'Modelling the Relation between Theology and Science'.

Markus Mühling dissects each of these topologies yet further, making the claim that interdisciplinary interaction itself is the ground over which we walk. In this way, scholars are challenged not to stake a claim in a particular territory, but to participate in the drawing of the map itself. Mühling introduces the neologism of 'wayformation' for this activity, and describes the interdisciplinary endeavour as one constantly swelling and reforming its landscape. One is reminded of Certeau's seminal essay 'Walking in the City', in which the walker can choose to leave the panoptic perspective and come down to street level, thus escaping the 'geographical space' of 'theoretical constructions' altogether (Certeau 1988, 93). Instead, the walker constructs and shapes the landscape as she passes through it. In this volume, however, our walker is not faced with the topology of a city, but of an entire universe. The shifting gaze of our contributors and readers moves not from the top of a skyscraper to the street, but from the telescopic to the microscopic and back. As Carl B. Pilcher so elegantly points out, these chapters each in their own way attempt to examine where we stand as human beings. Are we in among the atoms or overlooking the cosmos?

Certeau's description of the walker as one who creates the world around her (by choosing to actualize some possibilities and ignore others) reminds us that at heart these are ethical questions. Howard Smith, in his chapter 'The End of Copernican Mediocrity', uses the statistical unlikelihood of contact with intelligent life as a timely reminder of the ethical implications of our solitude in the universe. This sets a compelling framework for Erik Persson's chapter 'A Philosophical Outlook on Potential Conflicts between Planetary Protection, Astrobiology and Commercial use of Space', in which we are reminded of the metaphor of a footprint on unchartered territory, as he examines the ethical and moral implications for our interactions with microbial life on neighbouring planets. Zoë Lehmann Imfeld's chapter, meanwhile, uses science-fiction literature to explore the evolution of humankind as one changing the universe and being changed by it.

Indeed, this sense of changing and being changed is one which repeatedly emerges across our contributions, and is something that returns us to the sense of the landscape as a tangled web (to mix our metaphors). We are shown the position of humankind to be inextricable from its environment, and likewise the challenges and questions posed by other disciplines to be inescapable, either for the sciences or the humanities. Fabien Revol's chapter describes the interdependence of biodiversity, and answers both Persson and Smith by placing humans in an ecological relationship with our environment. For Bob White, our position as stewards of this delicate biodiversity offers potential for human beings in return, as a redemptive opportunity. This opportunity comes about, explains White, if we recognize the interaction of disciplines as a web of activity, unified in a common goal.

Intelligent design or indeed theologies of a 'God of the gaps' do not work because they are *neither* science *nor* theology. In 1990 Pope John Paul II wrote: 'Science can purify religion from error and superstition; religion can purify science from idolatry and false absolutes. Each can draw the other into a wider world, a world in which both can flourish' ('Message', M13). The contributions in this volume challenge and contest each other as often as they agree, but each one is attempting, at least, to draw a map which is not 'neither/nor', but 'both/and' – a map which has room for the marvels emerging from scientific discovery, and allows for the philosophical and theological questions that these discoveries raise. This is the map of the common cosmos through which this book travels.

References

Barbour, Ian G., *When Science Meets Religion* (New York: HarperOne, 2000).

Certeau, Michel de, 'Walking in the City', in *The Practice of Everyday Life*, trans. Steven Randall (Berkeley: University of California Press, 1988), 91–110.

Darwin, Charles, *On the Origin of the Species by Means of Natural Selection*, 2nd edn (London: John Murray, 1860).

Francis. 2015. *Laudato Si': On Care for Our Common Home* [Encyclical].

Gould, Stephen J., *Rocks of Ages: Science and Religion in the Fullness of Life* (New York: Ballantine, 2002).

John Paul II, 'Message if His Holiness John Paul II', in *John Paul II on Science and Religion: Reflections on the New View from Rome*, ed. Robert John Russell, William R. Stoeger S.J. and George V. Coyne S.J. (Vatican: Vatican Observatory, 1990).

Losch, Andreas, *What Is Life? On Earth and Beyond* (Cambridge: Cambridge University Press, 2017).

Losch, Andreas, 'The Need of an Ethics of Planetary Sustainability', *The International Journal of Astrobiology* (2018): 1–8.

Williams, Rowan, *Faith in the Public Square* (London: Bloomsbury, 2012).

World Commission on Environment and Development, *Our Common Future: Report of the World Commission on Environment and Development* (Oxford: Oxford University Press, 1987).

Part I

APPROACHES

Chapter 1

CONVERSATIONS ALONG THE WAY: HOW AND WHY SCIENCE AND THEOLOGY NEED TO INTERACT

Markus Mühling
Protestant University of Wuppertal/Bethel

Dialogue or Conversation?

Since the 1980s simple classifications of the relationship between the natural sciences and theology have mostly been abandoned. In their place a variety of descriptions of diverse types of dialogue between science and theology have been proposed, most prominently the process-thought inspired typology by Ian Barbour (i.e. conflict, independence, dialogue, integration), which has become a starting point of reference (Barbour 2000). Although these typologies have helped to overcome problematic dichotomies, it was felt that they had shortcomings in describing the reality of the relationship between theology and science, and many improvements have been suggested, such as a cyclical alteration between the particular types (Losch 2011, 88f.) or a simultaneous appearance of the particular types in different respects (Mühling 2014, 24). Could it be, however, that the main problem of this and similar typologies is simply that they are too abstract, and that this abstraction comes from the fact that the notion of 'dialogue' underlies all types? A dialogue is a talk between two speakers, who talk *across a subject matter* and who throw arguments and propositions at each other, pick up what they receive, alter it and return it like ping-pong players exchange balls across the table. This metaphor could just as well be a metaphor for conflict (which is simply a dialogue in which the players baulk at everything they receive), for independence (where the players are at different tables), and for integration (where the play across the table is only possible between partners who change their positions and are able to be at two locations at the same time). Fancy theories about dialogue have been elaborated not only as a basis for the description of the relationship between science and theology, but more broadly, to describe the work of human reason in general, such as Habermas's famous non-coercive discourse of the abstract, but ideal speech situation (Habermas 1983, 53–135).

What if the model of a 'dialogue' is simply not adequate to describe what is going on between science and theology? What if a 'dialogue', an exchange of *logoi dia*, is not able to depict reality? If 'dialogue' expresses an exchange of speech acts back and forth, in which the subject matter functions only as the means for an 'across' like the table in ping-pong, then this fails to express what is going on between science and theology, and it is not able to deliver the basis for the claim that the sciences and theology somehow need each other.

Standpoints or Perspectives on the Move?

The first problem with 'dialogue' is that it needs the abstraction of a standpoint in order to work. A standpoint is a fixed point of perception, argumentation and action, from which the players act. And, if it is necessary to leave this point for a while in order to get the arguments one wants to address, then it is also necessary to come back to this point for the next course of exchange, just like the ping-pong player who has a base position to which she returns after having lunged for the ball (see for instance Evers, this volume). But academic work – at least as long as one is doing research – cannot be depicted by certain moves back and forth from one standpoint, but is more like a constant movement into the unknown. Whether science or theology, research is not done from a standpoint, but by moving into the unknown. The concrete perspective from which all theories are made, all observations are performed and all arguments are exchanged is not a point of view, it is a perspective on the move. Neither is this constantly moving perspective a chain of connected points. Connected points (as visible in research articles) are only a very crude simplification of what is going on in everyday research.

Across or Along?

If it is true that the perspectives of observing, theorizing and arguing are on the move, there can be no table or field to serve as something which one talks across. The subject matter is not a means for the interaction, but is rather the very ground one is moving through in research. In perception on the move the terrain discloses itself only along the way one is walking. The observations, insights, discoveries, theories and arguments as expressions of research are formed by the way itself and the way is formed by walking *along, not across* it. Perspective is not only 'perspective on the move', it is also a wayformational[1]

1. The term 'wayformational' is a neologism that refers to ideas by Tim Ingold (2007; 2011, 141–76; 2015). The definition of it is thus: The equiprimordial formation or emergence of a dynamic perceiver and a dynamic way through reality by which dynamic perceptions that cannot be reduced to static standpoints appear. The relationship between way and perceiver is an internal and dynamic relation between perceiver and environment. It is therefore neither passive nor active, but resembles more the 'middle voice' in classical Greek: The perceiver or walker makes the way, but the way also forms the perceiver or walker.

perspective, that is a perspective that is formed by the way and that itself forms the way. In describing research as walking along a wayformational perspective, any radical constructivism and any naïve realism (like in neo-positivism) are unmasked as making the same mistake: They rely on the logic of depicting the environment from a particular standpoint and they both make the mistake of representationalism in very similar ways, either by projecting the pictures of the mind onto the screen of reality, or by picturing the environment on the black wall of the closed camera obscura of the mind (naïve realism). From a wayformational perspective however, the way belongs at the same time to the environment and to the walker; they are interwoven, entangled and resonate with one other.[2] The way is part of both, meaning, therefore, that the researcher herself is not independent of the subject matter she is researching. The relationship between researcher and subject matter is not an external one, but rather internal. 'Subject' and 'object' are at most simple abstractions from what is going on and they are in danger of becoming myths if the subject–object distinction is used as a basis for all knowing (see Mühling 2014, 33f.; MacCormac 1976).

Between or In-between?

Any conversation by two partners is not a dialogue *between* them, but rather a conversation *in-between* a way. Therefore, the different descriptions of both inter- and transdisciplinarity are not completely wrong, but they do have their limitations. The main limitation consists in the fact that they take the sure ground of specific disciplines as a standpoint, and try to transcend this somehow; either by talking across disciplines or by mixing up two formerly independent disciplines into a new one. To regard research as relying on a wayformational perspective, however, means being guided by the phenomena. Every academic discipline becomes subsidiary to the appearance of phenomena for practical reasons, and the history of academia, with the change and evolution of new disciplines, confirms that disciplines have no sacrosanct boundaries, be they physics, biology, philosophy or theology. They have their relative purpose as long as they are meaningful for joint research on the walk in-between the subject matter, but their boundaries are fuzzy and no methodology remains permanently fixed. Nevertheless, there are preliminary disciplines and approaches, and therefore one has to explain what the differences among them may be. This is no easy task since with regard to the subject matter of this chapter there are no concrete disciplines, but rather the natural sciences, or 'science' as a whole on the one hand, and theology as a whole on the other.

2. I coin the term wayformational or wayformational in order to express this dynamic internal relatedness with the environment.

Perceiving in Faith and Theology

So far we have used the concept of the wayformational perspective in order to describe academic work. But academia is not the example par excellence for perceiving from a wayformational perspective. The example 'par excellence' is life itself, and more concretely, faith. Whatever one does, perception comes first. One is immersed into perception by the whole living body, including its interbodiliness. In perception one perceives as one is perceived, driven completely by what is perceived. It was Merleau-Ponty who pronounced the unity of the perceiver and the perceived as *flesh* (1968, 248–51), where the perceiver is enrolled (Ingold 2015, 84–8) or immersed into the perceived. It is the world itself that perceives itself in the perceiver, but in some – not, of course, all – cases an asymmetry of perceived and perceiver is established in perception. In perception, there is no primary distinction between fact and value, but the perceiver responds to the affordances of the situations, ways or events of which she is a part (see Gibson 2015, 119–34). Perception is therefore always, in a specific regard, immediate. It is immediate because there is no primary need to voluntarily interpret what is perceived. The sun does not radiate, but shines, the stars do not beam but glow, driving a car is not simply an experience of acceleration, but excitement and so on. In another sense however, this immediate perception is at the same time mediated. It is mediated by the stories that the perceiver and perceived form – and by the attempt to retell these stories in human narrative. 'Stories are lived before they can be told' (MacIntyre 2007, 212). But there is a resonance between the stories that are lived and the stories that are told. This understanding of perception has two consequences: First, no one is alone in perception, but perception is always a communitarian activity. Second, perception is always embedded in stories and stories are always laden with meaning, but it is not necessarily the case that all meaningful stories we inhabit cohere to one another. Faith is nothing but the trust that despite all the incoherence and failure in perception, that in the end it is worth living a life in such a way. Christian faith is the trust that the Christian narratives provide a wayformational perspective of life that is able to incorporate all dichotomies and that leads to a reality-resonating mode of perception. Therefore, the Christian faith is not a realm separate from culture. One cannot distinguish between activities that are 'religious' and those that are not. It is not the case that performing a prayer is 'religious' or 'more religious' than meeting a fellow brother or sister during a walk on the streets. There are no specific 'religious' experiences like there are specific gastronomic experiences. If the expression 'religious experience' is meaningful at all, it has to be as an 'experience with experience' (see Ebeling 1975, 3–28; Mühling 2014, 86–96).

Perceiving in Christ by the Spirit is as realistic as perception in general. But a decisive sign of any claim to realism is the capacity for fallibility. It is precisely the fact that a perception can be wrong that makes it realistic. Perception can resonate the affordances of situations, but it can also be dissonant to the affordances of the world we inhabit. In both cases perception is, however, still mediated; in both cases perception appears as a disclosure of truth. Theology as an academic discipline is nothing but the methodically guided self-reflection on the conditions of perceiving along the Christian wayformational perspective. There is no contradiction therefore between Barthian-influenced descriptions of theology (or dogmatics) such as 'the scientific self-examination of the Christian Church with respect to the content of its distinctive talk about God' (Barth 1957, §1, 3) and Schleiermacherian descriptions as the knowledge of the necessary capabilities for the life of the Christian community.[3] Theology, in short, has to start with the everyday experience of the Christian wayformational perspective, trusting that this perspective is only possible thanks to the constant self-disclosure of the triune God, and it has to pronounce that on this wayformational perspective or mode of perception that there is no distinction between facts and value. Theology is committed to concrete life and perception, meaning it can grasp the general only via the particular. Therefore, any theology that would claim to not be a Christian theology, or a Jewish theology, or a neo-platonic theology or a biological theology or any other kind of particular theology would not be a theology at all.

Philosophy, Observation and the Sciences

Immediate perception is mediated by the stories we inhabit, but the stories we inhabit are also bodily stories. Our means of perception is our interbodiliness: skin, eyes, smell, our stomach and our nervous system belong to these means as well. If perception is always interbodily perception, one may ask, must there not be any common kind of perception, common to all wayformational perspectives and common to all modes of perception? The answer can only be that there is no such common perception, but that some commonalities can be (re)constructed or made visible by voluntary interpretation. One has only to have in mind that such activity does not provide a representation of reality, but rather an abstraction from reality. In this way, the interpretative abstractions are of immense value without producing false myths. And these kinds of interpretations are twofold. On the one hand, one can reflect interpretatively on the conditions of perception or becoming on a wayformational perspective as such – that is the task of the endeavour called philosophy. On the other hand, one can investigate concrete perceptions by means of observations and their interpretation – that is the task of the sciences. This task is of course not easy. One has to restrict oneself only to those features which are observable independent of the wayformational perspective; one has, in other words, to introduce the subject–object distinction, the ideal of objectivity, the idea of a separation between fact and value and so on. Nothing is wrong in this endeavour – as long as one is not claiming to gain a complete picture of reality by

3. Cf. §5 for the claim that the seemingly contradictory definitions are really only pronouncing different aspects of the same phenomena (see also Mühling 2012, 37–40). A detailed analysis of the similarities between Schleiermacher and Barth can be found in Stratis (2012).

it. It resembles somewhat that which has been called the 'natural attitude' (Husserl 1950, 63).

On this basis, one could now dig into the history of the theories of the sciences in order to describe more precisely this scientific endeavour. But the task of this chapter is to describe the conversation between science and theology from the wayformational perspective of theology perhaps a little more precisely than by using the metaphor of 'dialogue', thus making an excursion into the theory of the sciences unnecessary.

At this point we can see that any conversation between science and theology has to be in reality a conversation between three partners, science, theology and philosophy, independently of whether there is a professional philosopher involved or not.

When theologians, scientists and philosophers talk to each other in conversations, what kind of conversations can be distinguished? I want to suggest a very simple classification drawn from the model of possible conversations one could have while walking with others.

1. Following Conversations: In many situations during a walk, the terrain requires the walkers to follow one another. In this case, the first walker functions as a kind of guide, whose attention and whose story of formation allows a specific kind of attention and perception. He will show the followers some of the experiences, will explain the background stories of what is perceived. He will encourage the followers to make their own observations and the others will follow in his steps. From time to time it might be necessary for the guides to change; then the guide becomes a follower and vice versa.

What sounds natural during a walk seems to be out of fashion if one applies this model to the conversation between science, theology and philosophy. No philosopher today wants to follow theology, at least in order to avoid the suspicion of the old *ancilla theologiae* model. Most philosophers, except perhaps some of the now old-fashioned naturalists, want to follow natural scientists by hypostasizing scientific research results. Likewise many scientists would deny that it is meaningful to follow either theology or philosophy in their research. And the same seems to be true for the theologian: If he follows a particular philosophy, he would – at best – be accused of being creative or denying the foundations of biblical faith. If he follows science he would be accused of leaving his own terrain. Following each other within one's own discipline seems to be a bit more acceptable – at least as long as it does not last forever. Sometimes, however, it seems – after Kuhn – that to not slaughter one's academic forbears is regarded as a kind of intellectual suicide. Moreover, following the footsteps of others is often discredited in the intellectual sphere of liberal (post)modernity. Such relationships are pictured normally with the help of hierarchical models or models of suppression and liberation. But is slavery and exodus really the general pattern for academic work? Is not the model of going out for a walk where the terrain sometimes makes it necessary to follow one another a better model, one that comes closest to what is actually going on? Did not Faraday follow his Sandemanian religious commitments even in his experimental work and in his preference for field theories instead of atomic ones (also sometimes denying this fact, see Mühling 2011, 237–54)? Did not Einstein follow, in his progression from the special principle of relativity to the general, intuitions that where impossible without his reading Hume and Schopenhauer (84–105, 145–53)? Would not have Karl Barth's theological commitment to revelation have been impossible without following his brother Heinrich in the interpretation of epistemology (Grube 1998, 102–47)? We could mention endless examples of the same kind. What seems to be odd in the models of hierarchy, suppression and liberation becomes quite natural in the model of hiking – and far more common than normally admitted. I therefore started with the model of walkers following after one another, simply in order to liberate it from its negative connotations. The blind spot, however, in interdisciplinary, phenomena-oriented research lies not in the fact that no one is following anyone else, because this is happening independently of whether it is admitted or denied. The blind spot is that we have no methodology and no academic ethics for *how* to follow one another and how to follow one another through different disciplines. Following one another is not without its dangers, although these might have been overstressed in the past. But following one another also enables us to discover new ground, where one cannot go without following someone else. There is a lot of work to be done here in the future.

2. Accompanying conversations: Imagine a situation in which the terrain is wide enough to walk side by side. Neither of the partners walking and conversing can have the same impression of their environment simultaneously, but due to their different stories of personal formation they will perceive things in a different manner that allows them to see more things together than one would perceive alone or two would perceive when walking side by side silently. Accompanying conversations might be somehow similar to the old discourse model, but there is an important difference: Since the ground is constantly changing, the discussion cannot be guided only by the intentions of the partners or by attention to the other partner, but must be guided by their way through the ever-changing ground itself. The discussion has to be an attentional rather than an intentional conversation (see Ingold 2015, 133).

Depicting academic conversations in this model seems to be far easier than in the first model. In a time in which third-party funding is being provided more and more by non-academic institutions for inter- and transdisciplinary work, such conversation in the company of others appears be a new scholarly ideal. However, there is still a lot to do in order to find some meaningful guidelines for an academic ethics for such interdisciplinary discussions. For example, one has to avoid the danger of being purely result oriented. In this case, accompanying each other would be nothing but a means to attaining third-party funding – to satisfy the expectations of the sponsors, to finish a publication or to reach targets. It is crucial to remain attentive to phenomena as they appear, and it is crucial to remain conscious of the fact that no one will close the ever-enfolding ground of discovery.

3. *Meandering Conversation*: Imagine a long walk through difficult terrain. Imagine that it is not clear which appearance will guide a fruitful way of travel. Imagine that it is necessary that the partners partly follow each other, partly accompany each other, but that is also possible to go different ways to the right and the left in order to meet up again in the future, so that it is necessary to be a follower for a short stretch in order to move forward at all. Imagine that this procedure is necessary due to the shape of the terrain. And imagine the discussions that will happen during the periods of reunification. Will not all who have gone different ways give hints towards the common way further on? Will they not from time to time insist that their perceptions while walking alone provide the best guidance for the path ahead? Will they not from time to time feel forced to contradict each other? And will not also these contradictions be fruitful for the way further on? Of course this model of meandering conversation also has its risks. There is the risk of losing one another forever while walking separately and therefore of losing the joint endeavour. There is also the risk of insisting so much on one's own impressions that one cannot reconcile them with those of the others in a fruitful manner, putting the joint endeavour at risk.

It might be easiest to apply this image to interdisciplinary academic discussions. I hope that many will concur that the model of meandering conversations is the most comprehensive one of the three I have tried to introduce. However, there is a decisive difference between going out voluntarily for a joint walk in a terrain that requires meandering pathways, and academically meandering discussions. The second ones are not usually voluntary. If it is correct that perceiver and perceived are shaping each other during perception, if it is true that walker and way are forming each other, and if it is true that the appearance of phenomena and disciplines and their tools are constantly shaping each other, then the danger of separating forever without meeting again means losing both the phenomena and the subject matter. One might be tempted to explain situations like these with the remark that there is no world, or at least not one world. But losing the concept of one shared world means losing the concept of a world altogether. And losing the concept of the world – despite the fact that it might not be a sharp concept and it is not easy to define – means losing any resonance with as well as the joy of discovery. Separating forever without any contact is also not a possibility, not for theology nor for philosophy and the sciences. They might survive as doctrines or technologies or means of power, but the truth of philosophy is no more reducible to questions of power than science consists in technology and theology in teaching doctrines. They do not have to construct a common ground – their ongoing life as research is enabled by different appearances on the different wayformational perspectives of a common ground.

Conversations while Walking Together and Barbour's Abstraction

We started with reference to Barbour's typology of conflict, independence, dialogue and integration. We have seen that this typology is too abstract, and we have seen that this abstraction relies on the mistake of depicting the relationship between science, theology and philosophy as a dialogical ping-pong game. And I have suggested the model of conversation during a walk including the subsets of following conversations, accompanying conversations and meandering conversations as a better model. If this last claim is true, one would expect that the third model could incorporate the virtues of the first ones. This is indeed the case. The model of meandering conversations is not only able to include conversations that occur side by side and conversations that occur forwards and backwards, but it is also able to include conflict, independence, dialogue and integration. However, there can be no conversation that is purely adversarial. There can also be no permanent independence. Lastly, if reality consists in ever-changing alterations of processes and storied events, as is said by process-thinkers, any possible integration that might appear would be necessarily a transient one.

If one does not see academic endeavour in a quasi-platonic way that separates a possible 'essence' of a discipline from the researchers, their lives and their activities, it becomes important how one conceives inter- and transdisciplinary work. The models and pictures we use in a hidden way in order to structure the relationship shapes our modes of perception and our attitudes of behaving – in life as well as in academics. Barbour's typology is certainly able to give a certain frame for perception and it was decisive in overcoming false dichotomies, but at some point it also increases certain dangers; for example the danger of classifying oneself and one's conversation partners into one of the types, as well as the danger of conceiving research in a purely intentional manner. Conceiving research in the picture of conversations during an incompletable walk might be able to generate more fruitful attentional attitudes.

Why Is It Science, Theology and Philosophy that Need to Interact?

Christian theology is the self-explication of the concrete Christian wayformational perspective and it pronounces the inseparability of fact and value. The sciences rely on observations and their interpretations by theories abstracting from concrete perception. Nevertheless, they are immersed in the same changing ground and its appearances in perception. Without conversations shared by walking this ground together, they degenerate and lose their own identity.

Theology in this case would become the endeavour of interpreting doctrines, which is harmless as long it remains purely academic. But theology is related to a different public realm – not only the academic one, but also the civil society and the church, and that means contact with people in everyday life, their perception and their activities. If academic theology were to degenerate into the art of only interpreting doctrine, then in the life of the churches it could all too easily be seen as a means only for prescribing doctrine. Theology would then lose its own signature of being dependent on the grace of the triune God and his self-disclosure that leads humans into a Christian-storied way of perceiving. Someone who regards a doctrine as something to be prescribed also regards it as being at their disposal, and that is nothing but a contradiction of justification by grace. How can conversations with the sciences and philosophy be helpful in avoiding this danger? There are many good answers for this. I think the best one might be that we are bodily creatures and that the Son became also flesh. Faith, therefore, has nothing to do with following doctrines and acting intentionally according to them, but firstly with perceiving oneself as intermeshed into the world in the way of the stories of the Son and the Spirit. Incarnation and concarnation (which describes the activity of the Holy Spirit better than the notion of fleshless inspiration) mean that the abstract way of the sciences in interpreting observations *etsi deus non daretur* have to be recognized by theology as the self-interpretation of bodily perceivers on a Christian wayformational perspective.

The sciences, on the other hand, are in danger of degenerating either into technologies or becoming quasi-religions. The degeneration into technology occurs when scientists claim that the methodologically useful distinction between fact and value becomes a strict separation. In this case, they no longer see that curiosity is not only a decisive attitude of research, but also of care. Curiosity and care are different sides of the same coin. And talking to theologians and philosophers (both!) could be a means to prevent the danger of separating curiosity and care. Scientists have claimed too often that the results of their research are ethically neutral and that they can deliver the questions of applicability into other hands – the hands of theologians, philosophers and politicians! To give examples would be far too depressing.

In a pluralistic age many scientists might not go the route of degeneration into technology, but rather of becoming founders of krypto- or quasi-religions. This danger emerges when science is not aware that in its interpretation of observables it relies on an abstraction from perceived reality. Sometimes the language of 'facts and nothing but the facts' suggests that the sciences describe the real world representationally. We have observed this kind of degeneration in some branches of the neurosciences and of sociobiology, or in some aspects of post-humanism. Both dangers also emerge when scientists are in discussion with theologians and philosophers, but perhaps not as easily, because they can then partake in a joint discussion about the fact that not only what seem to be facts are guiding all of our action and behaviour, but also empirically non-testable presuppositions which emerge immediately in perception and which are bound to the wayformational perspectives of different faiths or religions. I am not sure whether the term 'religion' (in contrast to faith) really denotes anything meaningful. However, there is one (though perhaps historically incorrect) etymology, coined by Lactantius, that makes good sense: If religion comes from *re-ligare*, to re-tie or re-bind, then one can interpret religion as releasing us from our interpretations and naturalistic attitudes and binding us back to the perception of reality (see Ingold 2013, 734–52, note 13, 750).

Philosophy, finally, deals with the very basics of perception (epistemology in traditional terms), becoming (ontology in traditional terms) and the perception of values (ethics and aesthetics in traditional terms). As such philosophy faces a danger that could more easily emerge when the conversational bounds to theology and the sciences are cut off: The danger of constructing systems and the danger of reducing their truth to their being accepted. But this is also degeneration because it ultimately means reducing all philosophical questions to power: power of interpretation, power of acceptation and so on. In this case, philosophy degenerates into a provider of ideology for politics.

The dangers in our postmodern world are manifold, and it is impossible to mention even a few without being far too simplistic. Nevertheless I want to make some concluding remarks. The dangers of living unbound to reality, the dangers of perceiving in a non-reality-resonating way and acting according to these kinds of perceptions and the attitudes they generate seem to be obvious to me. Living purely intentionally in an individualist way that subordinates everything and everyone as a means for one's own wealth seems to be one of the most plausible, but least reality-resonating temptations of our time. This attitude of reckoning with humans as 'autonomous' individuals rather than as particular persons in communion, of reckoning with humans as beings rather than becomings, surely cannot be entirely thwarted by sharing conversations while the academic disciplines of theology, philosophy and the natural sciences walk along together. But a separation and an absolute independence of these disciplines from one another surely furthers these temptations more than it does to liberate us from them.

References

Barbour, Ian G., *When Science Meets Religion* (New York: HarperOne, 2000).

Barth, Karl, *Church Dogmatics* (Edinburgh: T&T Clark, 1957).

Ebeling, Gerhard, 'Die Klage über das Erfahrungsdefizit in der Theologie als Frage nach ihrer Sache', in *Wort und Glaube, Bd. 2: Beiträge zur Fundamentaltheologie*, ed. Gerhard Ebeling (Tübingen: J.C.B. More, 1975), 3–28.

Gibson, James Jerome, *The Ecological Approach to Visual Perception* (New York: Psychology Press, 2015).

Grube, Dirk-Martin, *Unbegründbarkeit Gottes?* (Marburg: Elwert, 1998).

Habermas, Jürgen, 'Diskursethik – Notizen zu einem Begründungsprogramm', in *Moralbewußtsein und kommunikatives Handeln*, ed. Jürgen Habermas (Frankfurt/M: Suhrkamp, 1983), 53–135.

Husserl, Edmund, *Ideen zu einer reinen Phänomenologie und phänomenologischen Philosophie 1* (Den Haag: M. Nijhoff, 1950).

Ingold, Tim, *Lines: A Brief History* (London: Routledge, 2007).

Ingold, Tim, 'A Storied World', in *Being Alive: Essays on Movement, Knowledge and Description*, ed. Tim Ingold (Oxon: Taylor and Francis, 2011), 141–76.

Ingold, Tim, 'Dreaming of Dragons: On the Imagination of Real Life', *Journal of the Royal Anthropological Institute* 19 (2013): 734–52.

Ingold, Tim, *The Life of Lines* (London: Routledge, 2015).

Losch, Andreas, *Jenseits der Konflikte: Eine konstruktiv-kritische Auseinandersetzung von Theologie und Naturwissenschaft* (Göttingen: Vandenhoeck & Ruprecht, 2011).

Maccormac, Earl R., *Metaphor and Myth in Science and Religion* (Durham, NC: Duke University Press, 1976).

MacIntyre, Alasdair, *After Virtue: A Study in Moral Theory* (London: Bloomsbury, 2007).
Merleau-Ponty, Maurice, *The Visible and the Invisible* (Evanston: Northwestern University Press, 1968).
Mühling, Markus, *Einstein und die Religion: Das Wechselverhältnis zwischen religiös-weltanschaulichen Gehalten und naturwissenschaftlicher Theoriebildung Albert Einsteins in seiner Entwicklung* (Göttingen: Vandenhoeck & Ruprecht, 2011).
Mühling, Markus, *Systematische Theologie: Ethik. Eine christliche Theorie vorzuziehenden Handelns* (Göttingen: Vandenhoeck & Ruprecht, 2012).
Mühling, Markus, *Resonances: Neurobiology, Evolution and Theology: Evolutionary Niche Construction, the Ecological Brain and Relational-Narrative Theology* (Göttingen: Vandenhoeck & Ruprecht, 2014).
Schleiermacher, Friedrich Daniel Ernst, *Kurze Darstellung des theologischen Studiums zum Behuf einleitender Vorlesungen (1811/1830)* (Berlin: Walter de Gruyter, 2002).
Stratis, Justin, *God's Being toward Fellowship: The Meaning of 'God is Love' in Dialogue with the Theologies of Friedrich Schleiermacher and Karl Barth* (PhD Thesis, University of Aberdeen, 2012).

Chapter 2

'GOOD FENCES MAKE GOOD NEIGHBOURS': WHY THE DIFFERENCES OF SCIENCE, RELIGION AND THEOLOGY MUST NOT BE BLURRED

Dirk Evers
Martin Luther University Halle-Wittenberg

There is no such thing as science, as there is no such thing as theology.[1] Both are abstract general terms, and it takes careful consideration not to misunderstand them as references to quasi-objective entities or even quasi-agents acting in history and culture. When we speak of science we refer to certain practices, communities, institutions and a whole variety of methods and bodies of knowledge, and at the same time we refer to discourses in which the label 'science' is used as a value-laden expression for certain ways to see reality. Discourses on science and religion regularly suffer from distortions because they compete for public attention and belief formation, and mix facts with values. When in the following I refer to science or the sciences, I hope that the context will clarify which aspect of what we call science I am referring to. More or less the same applies to the terms religion and theology.

As a term, religion in the modern sense was coined in early modernity. Its pre modern Latin meaning referred to appropriate worship, but not to sets of beliefs. Only with early enlightenment thinking did religion as a generic term become common, claiming that it refers to a natural kind of which all historical religions are exemplifications. Today, however, religion refers to very different kinds of phenomena, such as ritual practices, dietetic regulations, belief systems, social codes

1. With this fundamental thesis and with my metaphor of fences and territories I largely concur with, among others, Peter Harrison's view that 'science and religion are not natural kinds. ... Rather they are ways of conceptualizing certain human activities – ways that are peculiar to modern Western culture' (Harrison 2015, 194). However, I don't fully subscribe to his view of religion and science as turning from inner virtues to sets of cognitive thoughts and beliefs in modern times. This seems to be the case for the term 'religion', but somewhat loses sight of the cognitive and constructive efforts of early Christian as well as medieval thinking.

and so on from a Western perspective. As a loanword from Latin in most European languages, 'religion' as a concept usually lacks equivalents in languages outside the European context. It is by no means self-evident to what we refer when we speak of religion. Again, the term denotes certain practices, communities, institutions and very different sets of beliefs, and at the same time it is ubiquitous in discourses in which religion stands for certain ways to see reality, for example as open towards different notions of transcendence over and beyond the factual world. Religion must also be distinguished from theology, which itself can be part of religion insofar as it critically and constructively reflects the respective religion's truth claims from within its tradition, and builds a body of doctrine. But there are certainly religions that deny the importance of critical and constructive reflection with regard to their own tradition and practice, and there are forms of theology which have developed very critical attitudes towards traditional religious convictions. In its institutional form at universities or other institutions of higher learning it is again an invention of modern Western societies and has also taken different paths in different contexts.[2] And of course, since they are abstract general terms, religion, theology and science cannot interact as such – be it in conflict, dialogue or integration (see Barbour 1974). Only people can communicate and participate in public and academic discourses on scientific and religious world views, and they do so in very different situations, under very different premises and with very different goals. Therefore, if we want to reflect on the interaction and challenges between science and theology, with theology as the reflective form of religion, we have to consider carefully the differences between the fields and the different levels of possible discourses. I begin by explaining my understanding of science and theology.[3]

What Is Science?

Modern science has developed as a set of explanatory and theoretical disciplines. At their core, these disciplines try to develop formal, mathematical models of explanation of natural phenomena. Natural phenomena are those phenomena which can be identified and measured in space and time by human beings. Often these are experiments in a laboratory, but science can also refer to descriptions of, for example, planetary motion or the behaviour of primates in nature. In any case, scientific disciplines deal with objects and facts of reality in a third-person perspective, and they aim at describing those facts and objects by theories which

2. This is illustrated by a series of essays on the development of the dialogue between science and religion in different countries and cultures around the world, which Zygon published in 2015. See the editorial in Drees (2015, 151–4).

3. My vision of a clear distinction between science and theology/religion is by no means new or original. In my own way I try to argue in favour of concepts such as those of Alfred N. Whitehead (see Whitehead 1925), and of Stephen J. Gould, who invented the label 'Nonoverlapping Magisteria' (NOMA) for his view (Gould 1997, 16–22; 60–2).

ideally allow for mathematical modelling. In this respect, modern science is different from pre modern philosophy of nature, because it does not claim to understand the 'nature' of things, but limits itself to descriptions of functional relations between objects like relations of cause and effect. In short, science has moved away from concepts of substance and things towards concepts of measurement and function (see Cassirer 1923).

This methodological approach is now highly sophisticated and diversified, and it includes a whole set of skills and practices. Simple reconstructions of scientific method as they were given by the school of logical positivism, which reduced empirical method to the interplay of empirical observation providing facts with formal modelling of theory, must be considered incomplete and simplistic. Historical contingency, the community of scientists, semantics and metaphors, limits and peculiarities of formal models as well as pragmatic categories which Michael Polanyi referred to as personal knowledge have to be taken into account to get a fuller and more realistic picture of what we call science today.[4] We cannot go deeper into the philosophy of science, into scientific method, its critique and possible classifications of different scientific disciplines. A short list of different aspects must suffice here to get the flavour of the complexity and the overall direction of scientific method and scientific discourse:

- Science develops mathematical models of relations between empirically controllable parameters which allow for the prediction and manipulation of those parameters (laws of gravitation, laws of chemical binding, descriptions of developmental physiology, etc.);
- Science combines and differentiates these relations and infers further lemmas in order to develop larger and coherent bodies of theory for certain fields of phenomena and manipulation under different conditions (cinematics, organic chemistry, etc.);
- Science collects and classifies phenomena and relates them to theoretical bodies (geology – geochemistry, behaviour of insects – genetics – biochemistry, etc.);
- Science develops frameworks of empirically testable and mathematically describable models which allow for the coherent explanation and eventually prediction of historical phenomena (evolutionary biology, geology, cosmology, ecology);
- Science combines different theoretical models with expertise in technical engineering in order to develop technologies (space industry, chemical industry, medicine, pharmacy, computer technology, etc.).

One could add ever more scientific approaches to different phenomena. What they all have in common are two features: they try to develop models of the relations among certain phenomena which can be represented by functional mathematical

4. Polanyi emphasizes the role of the tacit dimension in knowledge, by which we know *how* to do things and which often exceeds knowing *what* we are doing (see Polanyi 2009).

models; and they put all these models to the test by empirical falsification and verification. In short, mathematical modelling and empirical testing lie at the heart of any scientific involvement with reality. Science does not map a given reality, but develops functional means of dealing with its (amazing!) reliability and algorithmic compressibility regarding certain of its aspects. Science tries to answer 'why' and 'how' questions with regard to empirically identifiable phenomena. Thus it tries to provide reliable and objective knowledge in the sense that it does not vary arbitrarily with different individual and subjective attitudes, feelings, moods, preferences, cultural settings and so on. Nonetheless it is deeply embedded into culture, language, history and other contingencies, and it is organized and pursued by a wide variety of individuals.

On a professional and academic level there is a growing diversity among the sciences today. Scientists know more and more about more and more narrow and specialized fields of research and expertise. The age of polymaths of science is definitely over, if there ever was any. And even those individuals who are celebrated as iconic figures that can speak for 'science' in general, as was the case with Albert Einstein and is the case, perhaps, with Stephen Hawking or Richard Dawkins, actually represent very narrow fields of scientific knowledge. No individual is able to assess the latest theories of, say, quantum physics, inflationary cosmology, genetics, neuroscience and artificial intelligence on the level of a well-informed researcher and integrate them into one coherent body of scientific knowledge. There is a need for transdisciplinary discourse and transdisciplinary translation *within* science. Biologists need to import expertise from computer sciences, brain researchers need to combine biochemistry, biology and physics with highly sophisticated technological equipment such as scanners for functional magnetic resonance imaging and so on. And ever more hybrid fields of science develop, which reach into historical, linguistic, psychological, sociological and other fields of academic disciplines usually not addressed as natural sciences. This even extends to religious studies. With the development of cognitive studies of religion, which explain the cognitive functioning of religious symbols and rites, science contributes to the interpretation of religious phenomena. Combined with evolutionary psychology, these interpretations are used to develop theories about the origin of religions, their historic development and their future fate, thus trying to turn religion into a scientific subject matter.

What is presented as the modern *scientific world view* in public discourses or, less rigorously, as a possible scientific account of reality does not follow directly from scientific knowledge or research, but comes in forms of integrative, generalized and simplified narratives that bring together quite heterogeneous aspects of general accounts of scientific knowledge on very different levels of corroboration and by means of tentative extrapolations.[5] I think it is safe to say that often in

5. A recent example is the bestselling books of the Israeli historian Yuval Noah Harari, who claims to ask very big questions and then answer them not by reference to religious authority but scientifically. For him that is a disjunctive alternative (see Harari 2015a, b).

present debates on science and religion it is not that scientific evidence stands against ignorant and naive religious prejudices, but that different narratives battle for recognition in public discourses. Thus it is as much a harsh oversimplification to contrast science and religion as two opposite, mutually exclusive, separate and block-like sets of convictions as it is to mingle them into one indiscriminate continuum of knowledge acquisition. There is a clear-cut, categorical boundary between scientific and religious engagements with reality as well as their theological reflection as will get clearer, I hope, with the next two sections.

What Is Theology?

Typically, theology is an academic discipline taught at universities, seminaries and other institutions of higher learning. Taken literally ('theo' as derived from the Greek term θεός = God or divine, 'logy' as derived from the Greek term λογία, meaning narration or critical account), the term refers to discourse on God or the divine. Consequently, Augustine of Hippo defined the Latin equivalent, *theologia*, as 'reflection or discourse on the divine'.[6] Since God or the divine is neither an entity in space and time nor an abstract formal function, the approaches towards God vary immensely, and they vary not only among different faith traditions, but also from individual to individual. And they vary in such a way that they do not form a coherent body of knowledge or disciplines. 'God' refers to very different concepts in different traditions.

I myself write from a perspective within a specific theological tradition. I engage in Christian theology as a Western Protestant theologian. Protestant theology takes its orientation from certain fundamental principles developed in the reformation period in the sixteenth century, which are often labelled with the fourfold *sola fide*, *sola scriptura*, *sola gratia* and *solus Christus*. These slogans stand for the specific focus of Protestant theology which is the gospel of God's salvific action in Jesus Christ, in which God has transformed and continues to transform the life of human beings towards the realization of faith, hope and love. Theology in this sense is the reflection on this ongoing process, and with its reflection it seeks to find adequate expressions for the meaning of Jesus Christ with reference to present interpretations of reality and human existence, for the understanding of human beings in their relation to God and to one another, and for shaping the Christian community accordingly.

Thus the starting point of Christian theology, at least in its Protestant version, is not metaphysical theory, but the reflection of the salvific activity of God in Jesus Christ and the ways in which human beings understand themselves as part of this activity. This was already the central focus in the Reformation era when, for example, Calvin stated that theological wisdom 'consists almost entirely of two parts: the knowledge of God and of ourselves' (Calvin 1559, 1, my translation), or

6. In *De Civitate* VIII, 1 Augustine speaks of theology as *de divinitate ratio sive sermo*.

when Luther defined the proper subject of theology in his explanation of Psalm 51 as 'the human being guilty of sin and condemned, and God the Justifier or Saviour' (Luther 1914, 328, my translation). Thus Christian theology begins with *faith seeking understanding*, and it critically deals with scripture and tradition in order to unfold the Christian faith and its implications in relation to modern interpretations of human existence, reality and knowledge. If it does so within the context of a university or other modern academic institutions, it must engage in discourse with other faith traditions and with secular, naturalist, agnostic and atheist views of human existence, reality and knowledge, and it must account for its own presuppositions.

It should be obvious that theology in this understanding is not and cannot be an explanatory reconstruction of certain phenomena of reality in a third-person perspective, but is always taking first-person perspectives (such as faith as basic trust and personal involvement) and second-person conditions (such as expressivist, non-designative aspects of language and semantics, cultural formations and hermeneutical approaches to verbal and non-verbal communication) into account. Theology as a reflection of faith does not aim at inducing or proving faith itself, although it is, of course, aiming at coherence and plausibility. God's existence, for example, cannot and needs not to be demonstrated (see Evers 2015), but its character, challenge and consequences must continuously be unfolded. Unlike science, which exercises strict methodological discipline in order to verify or falsify hypotheses of explanation of distinct phenomena in space and time, theology engages in open discourses which aim at integrating third- and first-person perspectives with critically reflected and elucidated second-person conditions. And its goal is not an explanation of regularities, which in some cases might allow for technical manipulation and utilization, but an exploration of reality and an orientation of human beings and communities towards participation in God's transforming presence.[7]

There are, of course, alternative understandings and ways of doing theology, some of which see science and theology as complementary partners which should closely interact and provide mutual support for their knowledge and theories, so that both intellectual enterprises together provide a fuller picture of reality. Thomas F. Torrance has argued for theological science, while Alister McGrath has written three volumes on scientific theology (Torrance 1969; McGrath 2001; 2002; 2003). And Wolfhart Pannenberg has developed theology as science of the divine ('Wissenschaft von Gott'), in which God as 'the all-determining divine reality could be measured against experienced reality' because theological 'assertions must be tested against reality' (Pannenberg 2008, 19, 21). My understanding of theology, however, is driven by the conviction that it does neither theology nor science good if both intellectual enterprises are not carefully distinguished and, to a large degree, kept apart. Any 'mixo-scientifico-theologia' or 'mixo-theologico-scientia'

7. Here I am indebted to and consider myself in close connection with Ingolf Dalferth's understanding of theology, (see for example Dalferth 2016; 2006).

is a helpless confusion which is trapped in futile and aporetic arguments about how to reconcile data and theories gained by scientific method with phenomena valid in a first-person perspective or with concepts important for human second-person relations. One can study these endless arguments for example in debates on human freedom over and against the findings of brain research or the concepts of meaning and teleology (see Nagel 2012). And they reoccur in debates about the question of whether or not cognitive studies of religious concepts prove or disprove their validity.

Fences: Science and Theology Distinguished

The proverbial title of this chapter is quoted twice in the well-known poem 'Mending Wall' by twentieth-century American poet Robert Frost (1874–1963). Despite its simple language and the rural scene it unfolds, Frost's poem shows a complex structure and addresses several themes at different layers, such as human fellowship and the role of boundaries in human society. One aspect of his description of human fellowship is given in the title of his poem and is reflected in the opinion and the behaviour of the neighbour with whom the narrator of this poem interacts. Every year they meet to rebuild the stone wall between their two farms. Thus the poem insinuates that fences have to be *reconstructed* and have to be agreed upon again and again. Transferred to our subject we can state that in changing historical constellations, with reference to different challenges and presuppositions and across different cultural formations, we must constantly distinguish empirical explanations from ways of engaging with reality, which are shaped by personal and inter-personal factors. Thus I suggest that a basic requirement for any fruitful exchange between science and theology is a careful distinction between their different engagements with reality and their relations to existential human questions. Only on an abstract level are science and theology two of a kind as human enterprises, but on any pragmatic and methodological level they are significantly and intrinsically different. Therefore, as a first step in developing a more comprehensive perspective on the possibilities and limits of any mutual exchange between scientists and theologians, I claim that both disciplines are specific engagements with reality based on certain practices of relating to different realms of phenomena.

I speak of different engagements with or different approaches towards reality rather than use the metaphor of different 'perspectives'. That metaphor is tempting, but also potentially misleading (see Dalferth and Stoellger (2004) for a collection of different perspectives on perspective, as well as Mühling in this volume). It is tempting, because we cannot take multiple perspectives at the same time, and thus this metaphor reflects that science and religion are different and mutually exclusive but at the same time might refer to the same reality and might be related to each other. Thus they could interact by asking: Why do you see what I cannot see and vice versa? How can I change my perspective so that I see things which you have already seen? Geometrically and topologically it is essential to take different perspectives in order to develop a more comprehensive picture of reality. And as in the case of binocular vision, two perspectives combined allow for adding the third dimension to two-dimensional perception. But because of this insinuation the metaphor of perspective is also misleading. It suggests that there is a given reality independent of both perspectives to which they jointly attend to. And it suggests that you can get a more comprehensive picture once you know how to combine both perspectives. All this is not the case if one deals with science and theology. At the heart of both disciplines are not different ways of mapping a given reality, but different ways of practice, different ways of actively getting involved with reality. And this explains why, for example, scientific knowledge has to be transformed into narratives in order to be integrated into a comprehensive view of reality.

The fundamental methodological difference between science and religion is such that they do not complement one another. The combination of their respective claims, theories and methodologies is not a step towards a more *complete* explanation of reality. They do not investigate different segments of reality, which in a synthetic theory on a higher level can be put together like pieces of a jigsaw puzzle, but they refer to reality in different ways. This implies that the different 'perspectives' on reality as developed in science and religion are not held together by complementary references *to* reality, but by the fact that they are both activities with which human beings are actively involved *in* the process of reality. As I have already suggested, the main point of difference between scientific approaches and religion as well as its theological reflection lies in science's strict limitation to empirical investigation of measurable states of affairs in a detached third-person perspective and to reconstruct these investigations with functional models and descriptions, whereas religion is a way of finding meaningful orientation for human existence by integrating objective factual knowledge with personal intuitions, intentions and reflections in a first-person perspective and expressing and actualizing these perspectives in second-person relations. Thus religion and theology as its reflective self-reference aim at discursive, critical and argumentative forms of communication. They address *people* and intend to have an impact on individual conducts of life and on communities at different social levels (families, congregations, societies). And since religion and theology in a Christian sense eventually aim at communicating the gospel of the presence of God's kingdom and want to participate in its coming, they also – on a fundamental level – must be understood as a form of prayer, so that they also – directly or implicitly – address God whom they want to serve and whom they cannot control or verify.[8] Theological accounts of human existence and reality at large are not detached descriptions and theoretical explanations of certain states of affairs, but are attempts to effectively and creatively express the Christian faith as a way of

8. See Barth (1963), ch. 14, for example: 'Theological work does not merely begin with prayer and is not merely accompanied by it; in its totality it is peculiar and characteristic of theology that it can be performed only in the act of prayer' (160).

exploring the splendour, the challenges and the dark sides of human existence and getting involved in establishing relations between human beings and the divine.

If this is true, then it also explains the often bemoaned asymmetry of exchange between science and theology, which actually is quite appropriate. Religion and theology must take means of objective experience, factual states of affairs and empirical, scientific knowledge into account if they don't want to become illusionary and unable to fulfil their communicative task. They would lose what Ian Ramsey described as the 'empirical fit' (Ramsey 1965, 59). Religious language as well as its theological reflection must in one way or the other 'incorporate both the facts and features of the world …, and something over and above those facts and features' (Ramsey 1973, 60).[9] And since today any valid and realistic understanding of human existence and reality at large is informed by scientific knowledge – for some in more detailed ways, for others rather superficially – religious world views are challenged by scientific knowledge and theology must engage in qualified debates with scientific research and method. However, while theology as an academic discipline today must include reflections on scientific knowledge and methodology, science does not have theological evaluation among its methods or subject matters.

Thus I argue for *rigorous and self-critical intersubjective science* which carefully applies strict methods of empirical corroboration and uses mathematical modelling for its engagement with those aspects of reality that can be measured and functionally explained. And it is self-critical insofar it allows for verbal accounts of its methods, theories and data that reflect the bearings, limits and constraints of its empirical underpinnings as well as its theoretical models and extrapolations. Thus it is part of its rigour to criticize misconceptions of science which attempt to force explanations in any field of human knowledge into models derived from one or another particular science. Rigorous science is relieved from demands to provide meaningful world views, and it is free to concentrate as closely as possible on functional explanations within certain methodological frameworks. Scientific abstinence from religious-like life-forming claims and a hesitance, by its own methods, to finally decide on religious, ideological or ethical matters is in itself a positive form of a respectful scientific attitude towards reality, which must be much more valued by representatives of religions as well as theologians. A rigorous, but non-ideological scientific attitude is the best friend of open, critical and constructive philosophical and theological discourse. Interest in nature and its phenomena simply for the sake of knowledge, the joy of discovery, the sense of awe of the mysteries of nature, which is increased rather than diminished with every unravelling discovery, the humility which often goes hand in hand with the interplay between falsification and success in research, the rigour and enlightening clarity of careful empirical and theoretical method, the striving

9. This is what Ramsey called 'cosmic disclosure' and what German hermeneutic theologians addressed as 'experience with experience' (see for example Ebeling 1975, 3–28; Webster 1986, 124–5).

towards undistorted contact with physical reality while leaving behind personal preferences, the great, revelation-like breakthroughs in the history of science – all that and much more is part of the grandeur of science and is in fundamental accordance with a consequent distinction between science and theology.

And I argue for *realistic theology* informed by science, but not for theological realism in the sense that theology develops descriptive concepts and models 'referring to the *reality* that is God and God's relation to humanity', so that these models 'depict reality' (Peacocke 1984, 40, 44). Theology must be realistic in the very broad sense that it respects referential aspects of reality which are to certain degrees independent of our conceptions of them, but it is scientific research that must develop adequate methods to identify such aspects. Even more, realistic theology presupposes that many of our claims about the world are true or false and not just epistemically successful or unsuccessful. According to Putnam (1978), realism involves three claims: (1) entities exist and matters of facts hold in reality independently of human beings (ontological claim); (2) truth is a semantic category, and when we make true claims about reality this is not a matter of convention, although we are often not in a position to verify or falsify respective propositions (semantic claim); (3) reality can be apprehended and understood in its regularities, some of its entities and fundamental structures (epistemological claim). However, this view also implies that results of scientific investigation of reality cannot be identified with God, and that on the other hand the notion of God does not improve the explanatory power of functional descriptions. There are no divine matters of fact or supernatural entities which science can identify or empirically verify. That does not imply that religious accounts of reality and their critical theological reflection are mere projections or only expressivist inner attitudes towards reality. A realist's rejection of theological realism does not necessarily imply theological anti-realism, but is in search of pluralist and pluriform perspectives on reality which allow for different and distinguished ways of getting in contact with it so that we may find meaningful ways to express the orientational relevance of faith for human existence.

This embraces the attitude that in scientific research we investigate reality *etsi deus non daretur* (as if God didn't exist), because if we want to avoid superstitious or magic concepts of the divine, God is not and cannot be a functional parameter among others, which can be pinned down by empirical method.[10] Neither is the

10. This formula is usually ascribed to Hugo Grotius, although it is not found explicitly in his work and the idea behind it can be traced back to late medieval times, when some theologians considered that it might be a property of a perfectly created world that it exists as if there were no God. Hugo Grotius justified the validity of natural ethical law with this idea. Natural law is valid 'though we should even grant, what without the greatest Wickedness cannot be granted, that there is no God, or that he takes no Care of human Affairs: *etiamsi daremus, quod sine summo scelere dari nequit, non esse Deum, aut non curari ab eo negotia humana*' (Grotius 1720, Prolegomena No. 11, English translation taken from Tuck 2005, 89).

distinction between God and creation a distinction we could draw within our experiential world by scientific means, nor do religious convictions necessarily facilitate scientific investigation. Religious convictions might be a source of inspiration, as it was the case with many naturalists of early modernity, or on the contrary they might block creative imagination and the courage to think along new ways.[11] And while the cognitive attractiveness and behavioural significance of certain religious concepts may become subject of scientific investigation, truth claims in connection with religious world views as such cannot in all cases be settled on scientific grounds, because they usually don't provide explanations of causes and conditions, but explications of meaning and significance. This 'reminds us that the word "God" does not work as a high-grade scientific word at all. It is *not* a "hypothesis". God-sentences do not belong to the logic of science' (Ramsey 1952, 9). By definition, religious convictions can never be part of scientific methodology and argument, although they may influence actual scientific research. Theology reflects on the significance and meaning of religious and other perspectives on reality which try to give meaning to human existence and provide means of orientation to individuals and communities with reference to God, and 'reference to God is not reference to an explanatory principle but to the focal point of ultimate orientation' (Dalferth 2017, 75). The divine is not given, either as an external referential entity or as an inner spiritual fact. God is not an identifiable, separable object. Scientific attention cannot be directed towards this object and then find anything out about it, because God cannot be separated from God's environment. God does not 'explain' anything, because in a theological perspective God is the ground of being (see Tillich 1973) and thus in a sense the ground of everything. This, however, also applies to other all-encompassing notions like world or reality, insofar as they want to conceptually identify the realm of literally everything that exists. Even the universe or the cosmos or any scientific theory of everything is always 'smaller' (not in a topological, but a categorical sense) than reality, because they already sketch reality in ways relative to certain methods of investigation (see Gabriel 2015). With all our methods, including scientific ones, we only explore certain provinces of 'everything' in specific perspectives. To start with, there are blind spots in any such approach, which refer to the conditions of the possibility or the intentions of the corresponding perspective. The question, for example, why cosmology fascinates cannot be answered by finding out more about the universe. As Thomas Nagel famously put it, there is no 'view from nowhere' (Nagel 1986), and an overview of the whole is impossible. And not even by this

11. It can be argued that that was the case with Albert Einstein. What he called his cosmic religion seems to have inspired him to develop his special and general theory of relativity, but on the other hand it seems to have blocked him in later debates on quantum theory, when he repeatedly stated that God does not play dice, so that in a letter dated from 15 April 1954 Wolfgang Pauli wrote to Max Born about Einstein's stance in the debate on quantum mechanics: 'I entirely agree with your opinion that Einstein has "got stuck in his metaphysics"' (Einstein, Born and Born 1971, 116. See also Evers 2006, 5–27).

property, that if God exists, then God exists in an all-encompassing manner, can God be identified, because then the term would lose its distinctiveness and would be indistinguishable from terms like reality, world or nature.

But just as the meaning of human life arises out of our pluriform and orientational engagement with infinite reality, in which we are fortunately able to participate, so religious and theological approaches to reality claim that by doing so we also participate in God, who is present always and everywhere and who communicates in complex ways with human beings – to use a phrase of Lutheran eucharistic theology – *in*, *with* and *under* the conditions of finite reality. With reference to knowledge of God, we are dependent on God's specific involvement with reality. That is what I would call revelation, and that is what makes theology possible, not the manifestation of super-scientific facts, the superiority of a better explanation or the tentative successful testing of the implications of religious truth claims. However, in order not to become a meaningless, futile and illusionary enterprise, theology not only has to critically deal with its own religious traditions and religious codes of conduct, but must be apt to relate to different human engagements with reality including science as an important set of tools to understand and shape our reality.

This includes the conviction that the developments of modern science with their challenges for religious world views, traditional theological concepts and the debates that go along with them to this very day, are in the long run helpful for religion and theology to identify their own specific involvement with reality in the perspective of faith. Such developments prevent religion from turning into fundamentalist ideology, and are helpful for science to accept that there is no such thing as an undisputed scientific world view of everything. While well-defined scientific problems can be solved, philosophical as well as theological discourses are in principle endless. There is no final explanation. With regard to orientational knowledge including religious and secular world views we interpret explanations and relate them to broader interpretations of human existence, and we do so against the changing background of contingent historical situations and across different cultural settings. The overall framework of human life form and practice, which in modernity differentiated into scientific third-person perspectives, a rich variety of subjective first-person attitudes and different narratives, world views and ideologies, is a self-referential structure. It cannot be justified by further explanations and cannot be traced back to foundational ultimate reasons. The challenge is that there is no *final* explanation. That is what sets science and religion apart and at the same times makes it indispensable for any critical and constructive theology to relate to scientific knowledge at all levels.

Science and Theology as Neighbours

The above-mentioned poem by Robert Frost to which the title of this chapter refers begins with the line: 'Something there is that doesn't love a wall', and repeats this line in the final part of the poem. And indeed, this applies to our debate on science, religion and theology as well: there is something – throughout history as well as in present debates – that does not love the clear-cut and professional separation which we just discussed. We are all human beings, and every kind of knowledge tends towards views and understandings of reality which integrate facts and meaning. There is no *neutral* ground, but there is *common* ground beyond our disciplinary territories. This common ground is established by language, by narratives, metaphors, practices, media and institutions (like churches and academic institutes of higher learning), where processes of mutual exchange and of belief formation take place. Here the mutuality of exchange is essential.

As we have seen, religion cannot be separated from facts and empirical evidence in the same way that science must be separated from religious hegemony and interests. But with all its methodological discipline and theoretical rigour, science is still a human enterprise trying to explain different aspects of reality by getting into empirically controlled contact with them. It naturally tends towards shaping belief systems, and it naturally tends towards getting hold of religion as well as of other cultural phenomena in scientific perspectives. Sociology, cognitive science and neuroscience of religion, general religious studies, empirical anthropology, the economics and politics of religion and spiritual experiences – these and other disciplines continue to explore ways into the realms of religious beliefs and rites of individuals and communities (as an example of such discourse see Klein 2011). Today these investigations contribute to informed ways of discussing religions and their impact on people's world views, and rightly so. However, attempts by science to replace religion and to substitute theological discourse from within a certain religious tradition with allegedly neutral scientific studies may prove detrimental to science itself, because it does not account for the non-scientific presuppositions which it employs. Here again discourse and mutual exchange may be helpful also for science, when it helps to relieve science from excessive demands and from the temptation to stealthily ally with politics and power.

I have already pointed to some aspects of border traffic between science and theology (see also Polkinghorne 2000). Although good neighbours do not interfere in the other's territory and business, they meet at the border in order to mend or realign it, and sometimes they fight over it. Well-kept fences do not settle conflicts once and for all, but help to allow for a tense but productive interplay of cooperation and conflict, of agreement and dissent. Neighbours visit each other and get shown around the other's territory. They meet outside their own territory, on the market, in the pub, in church or in politics. Thus fences mark, separate and protect territories, but they do not isolate them. By separation and division they also establish relations. It is through boundaries and through negotiating them that we become neighbours. And while fences are intended to prevent assaults, hegemony and intrusion, they also establish neighbourhoods and exchange across borders. And when neighbours leave their territory, they stay neighbours. If science and religion meet *outside* their disciplinary territories, for instance in the academy or in the broader public realm, reflecting their respective contributions to our interpretations of reality, and if they are aware of their respective methodological and disciplinary rigour and limitations, then they will be able to address one another as neighbours and cultivate comprehensive views of reality by retelling narratives and establishing language games which are effective in expressing meaningful perspectives on human existence and which are appealing to people across disciplines, although the individuals involved may be far from total agreement and will never be able to provide a final explanation.

References

Barbour, Ian G., *Myths, Models and Paradigms: A Comparative Study in Science and Religion* (New York, NY: Harper and Row, 1974).

Barth, Karl, *Evangelical Theology: An Introduction* (Grand Rapids, MI: Eerdmans, 1963), 159–70.

Calvin, John, *Institutio Christianae Religionis* (Geneva, 1559).

Cassirer, Ernst, *Substance and Form and Einstein's Theory of Relativity* (Chicago, IL: Open Court Publishing Co., 1923) [German original published in 1910].

Dalferth, Ingolf U., and Philipp Stoellger (eds), *Wahrheit in Perspektiven. Probleme einer offenen Konstellation* (RPT 14) (Tübingen: Mohr Siebeck, 2004).

Dalferth, Ingolf U., *Becoming Present: An Inquiry into the Christian Sense of the Presence of God* (Studies in Philosophical Theology, 20) (Leuven: Peeters, 2006).

Dalferth, Ingolf U., *Radical Theology: An Essay on Faith and Theology in the Twenty-First Century* (Minneapolis: Fortress Press, 2016).

Dalferth, Ingolf U., 'A Relaxed View of Metaphysics: Neo-Aristotelian Thomism and the Theological Legacy of Thomas Aquinas', *ZThK* 114, no. 1 (2017): 49–81.

Drees, Willem B., 'Glocalization. Religion and Science Around the World', *Zygon* 50, no. 1 (2015): 151–4.

Ebeling, Gerhard, 'Die Klage über das Erfahrungsdefizit in der Theologie als Frage nach ihrer Sache', in *Wort und Glaube vol. III. Beiträge zur Fundamentaltheologie, Soteriologie und Ekklesiologie* (Tübingen: J.C.B. Mohr, 1975).

Einstein, Albert, Max Born and Hedwig Born, *The Born-Einstein Letters. Correspondence between Albert Einstein and Max and Hedwig Born from 1916–1955, with Commentaries by Max Born* (London: MacMillan, 1971).

Evers, Dirk, '"Der Alte würfelt nicht ...": Einstein und die Religion', *EvTh* 66, no. 1 (2006).

Evers, Dirk, 'Kann man Gott wissenschaftlich beweisen?', in *Wissenschaft und die Frage nach Gott. Theologie und Naturwissenschaft im Dialog*, ed. Andreas Losch and Frank, Vogelsang Frank (Bonn: Evangelische Akademie im Rheinland, 2015), 150–9.

Gabriel, Markus, *Why the World Does Not Exist* (Cambridge: Wiley, 2015).

Gould, Stephen J., 'Nonoverlapping Magisteria', *Natural History* 106, no. 2 (March 1997): 16–22; 60–2.

Grotius, Hugo, *De jure belli ac pacis, libri tres*, ed. Johannes F. Gronovius (Amsterdam: Wetstenius, 1720), X.

Harrison, Peter, *The Territories of Science and Religion* (Chicago: University of Chicago Press, 2015).

Harari, Yuval N., *Sapiens: A Brief History of Humankind* (New York, NY: Random House, 2015a).

Harari, Yuval N., *Homo Deus: A Brief History of Tomorrow* (London: Random House, 2015b).

Klein, Rebekka A., *Sociality as the Human Condition: Anthropology in Economic, Philosophical* and *Theological Perspective* (PSSR 3) (Leiden and Boston: Brill, 2011).
Luther, Martin, 'Enarratio Psalmi LI' (1532.1538), in *D. Martin Luthers Werke vol. 40/II*, ed. Karl Dreschner (Weimar: Hermann Böhlaus, 1914), 313–470.
McGrath, Alister E., *A Scientific Theology. Vol. 1: Nature* (Grand Rapids, MI: Eerdmans, 2001).
McGrath, Alister E., *A Scientific Theology. Vol. 2: Reality* (Grand Rapids, MI: Eerdmans, 2002).
McGrath, Alister E., *A Scientific Theology. Vol. 3: Theory* (Grand Rapids, MI: Eerdmans, 2003).
Nagel, Thomas, *The View From Nowhere* (Oxford: Oxford University Press, 1986).
Nagel, Thomas, *Mind and Cosmos: Why the Materialist Neo-Darwinian Conception of Nature Is Almost Certainly False* (Oxford: Oxford University Press, 2012).
Pannenberg, Wolfhart, 'Is There Any Truth in God-Talk?', in *The Historicity of Nature: Essays on Science and Theology*, ed. Niels H. Gregersen (West Conshohocken, PA: Templeton Foundation Press, 2008), 11–22.
Peacocke, Arthur, *Intimations of Reality* (Notre Dame: DePauw University, 1984).
Polanyi, Michael, *Personal Knowledge. Towards a Post-Critical Philosophy* (Chicago: Routledge, 2009).
Polkinghorne, John C., *Traffic in Truth: Exchanges between Science and Theology* (Norwich: Fortress Press, 2000).
Putnam, Hilary, *Meaning and the Moral Sciences* (London and Boston: Routledge & Kegan Paul, 1978).
Ramsey, Ian T., *Miracles: An Exercise in Logical Mapwork* (Oxford: Clarendon Press, 1952).
Ramsey, Ian T., *Christian Discourse: Some Logical Explanations* (London: Oxford University Press, 1965).
Ramsey, Ian T., *Models for Divine Activity* (London: S.C.M. Press, 1973).
Tillich, Paul, *Systematic Theology vol. I* (Chicago: University of Chicago Press, 1973).
Torrance, Thomas F., *Theological Science* (Oxford: Oxford University Press, 1969).
Tuck, Richard, *Hugo Grotius, The Rights of War and Peace Book I* (Indianapolis: Liberty Fund, 2005).
Webster, John, *Eberhard Jüngel: An Introduction to His Theology* (Cambridge: Cambridge University Press, 1986).
Whitehead, Alfred North, 'Religion and Science', *Atlantic Monthly* 136, no. 8 (1925), 200–7.

Chapter 3

MODELLING THE RELATION BETWEEN THEOLOGY AND SCIENCE

Andreas Losch
University of Bern

How can we imagine the relationship between theology and science? Is there, actually, any relationship between two such divergent disciplines? In this chapter I will try to sketch the modern assessment of the potential for interaction between these territories, elaborating on a complex pathway model proposed by Robert John Russell. While the chapter is admittedly more technical in style, it is important to employ a rigorous examination of such a delicate relationship as that between theology and science. In this way, we can avoid the pitfalls often seen when religious believers look to use science or interpret the history of science only in a way that supports their religious beliefs, or in a way which creates a story to tell. Likewise the historical development of the scientific method repeatedly demonstrates that there is no need for theological explanations in science, moreover, that those attempts only hinder the advancement of science. Famously telling Napoleon, 'I have no need of this [God] hypothesis', Laplace claimed that reflecting on God within mechanics led nowhere except to scientifically irresponsible suspensions of questions (see Henrich 2010, 166). Although it is a matter of dispute whether this anecdote only implies a necessary methodological atheism (or naturalism) in scientific practice or if it is also expressing personal atheistic convictions of Laplace (Henrich 2010, 168), the bottom line regarding science is clear: There is no need for God in the scientific method. The naturalism essential to science prohibits any reception of theology within the method of science.

For theology, the situation regarding science is different. While there is consensus that one must avoid a 'God of the Gaps' that would soon be undermined by scientific progress (cf. Losch 2014, 147), it is disputed whether one should include scientific findings in one's own theology of creation or not. Karl Barth decided not to do so, although he admitted that 'future workers in the field of the Christian doctrine of creation will find many problems worth pondering in defining the point and manner of this twofold boundary' between science and theology (Barth 1958, preface). Despite this confession, Barth's disciples had a hard time trying to motivate Barth to engage in a dialogue with science (see Clicqué 2001). On the other hand, Barth's Scottish follower Thomas F. Torrance was personally deeply engaged in dialogue with the sciences. Very generally, while the continental European theological tradition tends to, more or less, ignore the impact of the natural sciences regarding theology, there are strands within the Anglo-American tradition that are keener to enter into dialogue with the sciences. Crucial here is the old tradition of natural theology in the UK and particularly Ian G. Barbour's *Issues in Science and Religion* (1966) as causing a change of attitudes in the United States and beyond (see also Barbour 2008).

I sympathize with the Anglo-American approach, because although the traditional continental European segregation seems to be a peaceful solution, the dominance of the popular myth of a perennial conflict between science and theology (Harrison 2015) and the preferred media coverage of conflicts of science with creationism, for instance, result in misinformation to the public regarding the relation of science and religion. It is thus necessary, also in a European context, to actively approach the dialogue between theology and science, namely incorporating fundamental scientific findings such as evolution or Big Bang cosmology into one's theological concepts to protect believers from stepping into the trap of creationism. Like John Polkinghorne and Michael Welker, I assume that science and theology are partners in the pursuit of truth.

So, while science may not need to relate to theology, theology should relate to science. Their relation therefore is essentially *asymmetrical*, which shall be the first result of our attempt to model the relation between theology and science.

This differentiation presupposes an essential *separation* between both disciplines. It makes no sense discussing the opportunities for dialogue between science and religion while assuming one domain should be able to conquer the other, be it via creationism suppressing scientific insights about evolution or be it via scientism declaring any non-science to be explained by future science or even worse to be irrelevant. Nevertheless, a strict separation is of course another dead end, because no discipline would have anything to say to the other.[1] The common distinction that science is about 'How?' questions while theology answers 'Why?' questions I regard as a helpful start for a dialogue, because in this way each partner in dialogue takes the opponent seriously at least. It will

1. 'Although such an "independence model" … certainly presents a more healthy approach for theology than the idiocies of creationism, it could also result in an outdated imagination of the current status of scientific findings, which could, in turn, influence the implicit scientific assumptions of modern theologies in an anachronistic way' (Losch 2016).

not fulfil our aim to rest with this distinction, however.[2] As John Polkinghorne puts it, 'The answers to "How?" and to "Why?" must fit together without strain' (Polkinghorne 1998, 22).

This also means for Polkinghorne, a physicist, that 'physics constrains metaphysics' (Polkinghorne 2003, 34–5). Does science condition any philosophical or theological thought about nature in this way? On the one hand, this could indeed be true, reflecting the asymmetry mentioned between science and theology. On the other hand, there remains the question whether it would really be simply physics entering into dialogue with philosophy and theology or whether it will not always be an already metaphysically prepared *interpretation* of physics, namely some sort of *physicalism*, which seeks to impose its popular philosophical perspective on philosophy and theology proper. In the latter case, Polkinghorne's approach could then result in a physicalist-flavoured metaphysical dictate on theology.[3] With this caution in mind, how could and how should science and theology relate to each other in detail?

Science, Hermeneutically Conceived

It was Ian Barbour's pioneering approach to postulate a *parallelism* in the methods of science and religion, which sparked a complete renewal of the Anglo-American science and religion discussion. His thesis may, however, initially seem surprising. Isn't science about facts, objective observation and data-analysis, while religion is about personal faith?

Let us look first at how Barbour portrays science. The key to understanding his thesis is that he regards science – similar to humanities – as a *hermeneutical* enterprise (see Clayton 1989), meaning that also in science, interpretation is needed. Given what I call his 'spectrum thesis' of a *continuum* between the natural, social and human sciences (Losch 2010), this sounds conceivable to me. But is it conceivable to a natural scientist? In a Baconian view of science as a process of *induction* from observations and experimental data to concepts and theories, there would be no need for hermeneutics at all. Barbour, however, disregards the inductive view as inadequate 'because theories involve novel concepts and hypotheses not found in the data, and they often refer to entities and relationships that are not directly observable' (Barbour 1997, 106). Barbour's graph illustrating the structure of science shows two detours on the road between theories and observation/data:

2. Perhaps Barbour's fourfold typology of conflict, independence, dialogue and integration can also be imagined as phases of a dialogue (Losch 2011, 213).

3. For instance, Astrid Dinter worries about Polkinghorne's 'conditions of consonance' (Dinter 1999).

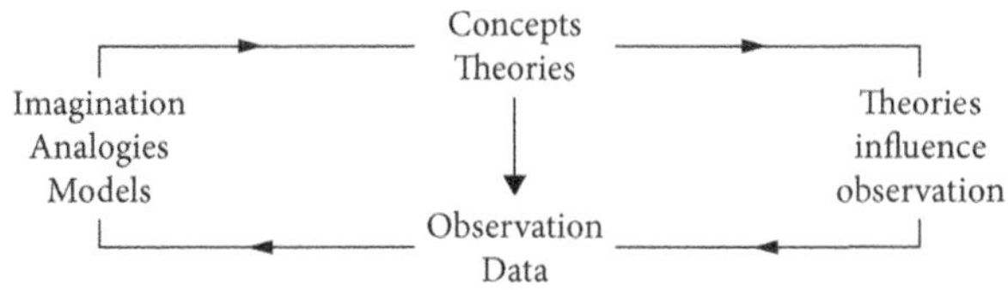

Figure 3.1 The structure of science.

'There is, then, no direct upward line of logical reasoning from data to theories in the diagram, but only the indirect line at the left, representing acts of *creative imagination* for which no rules can be given' (107). Analogies come into play, and frequently they are developed as a '*conceptual model* of a postulated entity that cannot be directly observed', which finally leads to the formulation of a generalized and abstract theory (107). Barbour illustrates his assumptions with the billiard-ball model of a gas from which the kinetic theory of gases was developed and which 'postulated invisible gas particles that were imagined to collide and bounce off each other like billiard balls' (107).

Still, there is the need for experimental proof as a key component of science. With the acceptance of this *hypothetico-deductive* view of science, Barbour claims to join forces with Hempel and Popper (340), so there is a *downward arrow* from theory to observation in his illustration. Following Popper, a disagreement with data would result in a *falsification* of theory.

Barbour is, however, critical of this sort of simplification, as he is concerned with not only philosophical abstractions about the scientific process, but concrete examples of the history of science as well. 'In some cases, discordant data were brought into harmony with a theoretical prediction by the introduction of *ad hoc auxiliary hypotheses* (107).' Technically, this appears to present an echo of Lakatos' version of philosophy of science. The historical examples Barbour provides, however, also seem logically coherent.

> Early opponents of Copernican astronomy said the hypothesis that the earth moves around the sun must be false because there is no visible annual change in the apparent position of near stars relative to distant stars. But Copernicus dismissed this discrepancy by introducing the hypothesis (for which there was then no independent evidence) that all the stars are very distant compared to the size of the solar system (107).

An auxiliary hypothesis saves his main assumption, therefore.

Also, Barbour's point here is the meaning of *coherence*. 'We can never test a theory alone, but only as part of a *network of theories*. If a theory fits poorly with the data at one point, other parts of the network can usually be adjusted to improve the fit. Theories with terms far from the observational boundaries are not uniquely determined by the data' (108).[4] This leads Barbour to the more general assumption

4. Barbour refers here to Willard van Orman Quine (1963).

that '*all data are theory-laden*. There simply is no theory-free observational language' (108), and this influence of theories on observation is shown on the right-hand loop of the diagram.

There is therefore also a necessary personal contribution within the very process of science. At least this was scientist-philosopher Michael Polanyi's point, to whom Barbour refers early in his writing (Barbour 1960). Barbour picks this up in framing his understanding of the process of science as *critical realism*. In his *Issues in Science and Religion*, Barbour defines critical realism (in partly Whiteheadian language) as that it 'must acknowledge both the creativity of man's mind, and the existence of patterns in events that are not created by man's mind. *Critical realism* acknowledges the indirectness of reference *and* the realistic intent of language as used in the scientific community' (Barbour 1966, 172). Also in *Religion and Science*, the book we have mostly closely followed here, Barbour labels himself a *critical realist*. There, he reframes his understanding of the term that it is *realism* because in his conclusion the meaning of truth is indeed correspondence with reality (an equivalent of the idea of the *agreement of theories with data*), yet the criteria of truth must include *all* aspects for assessing theories in scientific research. So, not only agreement with data, but also – and here it becomes clear why it is a *critical* realism Barbour advocates – *coherence, scope and fertility* (which resonate well with the coherent and pragmatic views of truth) (Barbour 1997, 110).

This all, however, applies to 'normal' scientific progress only. Barbour also supports Thomas Kuhn's idea of paradigm shifts in scientific revolutions, while being aware that 'here we see a contextualism, a historicism, and a relativism contrasting the formalism and the empiricism of Popper's account' (Barbour 1997, 109). For him, Kuhn's thesis includes a more pragmatic element of problem-solving, which is reflected in Barbour's criterion of fertility.

Barbour's hermeneutical interpretation of the structure of science appears convincing to me. Any positivist opposition against it ('science is only about facts') to me seems to result from a conflation of the ideal of purely objective science with the real process of scientific examinations. In particular, such a position ignores the inescapable reality that we, being part of nature, have no 'God's-eye view' with which to examine it; hence, a subjective or personal factor in science can never totally be ruled out.

Religion, Similar to Science?

Barbour shows himself initially to be deeply influenced by famous Anglo-American philosopher Alfred North Whitehead, one of whose statements he even quotes twice early on: 'The dogmas of religion are the attempts to formulate in precise terms the truths disclosed in the religious experience of mankind. In exactly the same way the dogmas of physical science are the attempts to formulate in precise terms the truths disclosed in the sense-perception of mankind' (Whitehead 1926, 57; Barbour 1966, 129). Following this quotation, one can better understand the assumption of a parallelism between the methods of science and religion, although it may still seem surprising. Such a parallelism is what Barbour tries to develop by stretching his idea of critical realism into the realm of religion as well. Polanyi may again have been of help here, having identified science as another sort of 'faith seeking understanding' (Polanyi 1946). As a result, the graph Barbour offers showing the structure of religion is very similar to the previous one showing a hermeneutically conceived science.

The downward arrow in this graph is shown as a dashed line, because Barbour is aware that 'the experiential testing of *religious beliefs* is problematic', although he does claim that there are 'criteria for judging the adequacy of beliefs' (Barbour 1997, 111) as well. More important is the loop on the right of the diagram, showing the influence of religious belief on experience and on the interpretation of traditional stories and rituals. Barbour, however, is essentially convinced that 'assessment of beliefs *within a paradigm community* can be undertaken with the same criteria listed above for scientific theories, though the criteria will have to be applied somewhat differently' (Barbour 1997, 113). The criteria are, again, *agreement with data, coherence, scope and fertility*. What is meant by these in the context of religion?

The 'primary *data*' are individual religious experience and communal story and ritual, for which a faithful rendition is required. As these data are much more 'theory laden' than in the case of science, one has to examine the influence of beliefs on experience and on the interpretation of story and ritual (Barbour 1997, 113). *Coherent* consistency with other accepted beliefs ensures the continuity of a paradigm tradition, although there is room for reformulation and interpretation. *Scope* is of importance as well, as 'religious beliefs can be extended to interpret other kinds of human experience beyond the primary data, particularly other aspects of our personal and social lives' (Barbour 1997, 113). In an age of science, this should include scientific findings. *Fertility* in religion has many dimensions. 'At the personal level, religious beliefs can be judged by their power to effect personal transformation and the integration of personality' (Barbour 1997, 113).

Robert John Russell called this Barbour's 'crucial, methodological claim that "bridges" science and religion' (Russell 2004, 49–59), and as we shall see Ian Barbour is convinced that 'the basic structure of religion is similar to that of

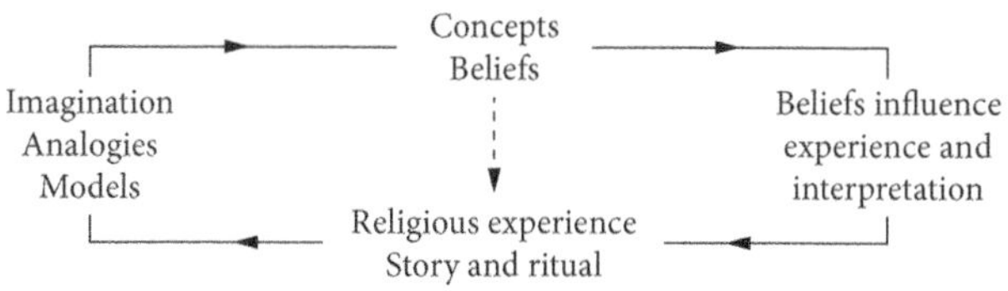

Figure 3.2 The structure of religion.

science in some respects, though it differs at several crucial points' (Barbour 1997, 110). I will offer my critique of the assumed parallelism within the shared concept of critical realism at the end of this chapter. For now, we will embark on discussing Robert John Russell's version of Barbour's findings, but for that, we have to discuss the idea of a hierarchy of the sciences.

A Hierarchy of Sciences?

The hierarchy of the sciences is a prerequisite to understanding Russell's arrangement of his pathway model. It is influenced in this regard by Arthur Peacocke's elaboration of this hierarchy, which in turn is inspired by Michael Polanyi (see Peacocke 1979, 371),[5] and develops thoughts of Bechtel/Abrahamsen (Bechtel and Abrahamsen 1991, 257). Against all sorts of 'nothing-buttery', which reduce every phenomenon to the lowest level of existence (for example, 'life is nothing but physics and chemistry'),[6] Peacocke favoured not only a hierarchy within the existing disciplines, but also a hierarchy of the sciences themselves (Peacocke 1993, 216–17).

Bechtel and Abrahamsen distinguish four levels of reality: the physical sciences (level 1), the biological sciences (level 2), the behavioural sciences (level 3) and the cultural products, arts and sciences (level 4) (see the graph in Bechtel and Abrahamsen 1991, 257). All levels are entangled through upward and downward arrows, showing those portions of the respective level of interaction with the next level. For instance, neurophysiology is imagined as influencing the behavioural sciences (upward arrow), which in turn are portrayed to influence behavioural biology (downward arrow).

Arthur Peacocke developed this rough draft into a much more detailed hierarchy of sciences, including horizontal hierarchies as well. For instance, on Peacocke's lowest level, elementary particles make up atoms, which make up molecules, which make up minerals, which make up planets (see Peacocke's graph in Peters 1998, 208).

Is this only a reflection of the ancient assumption of a *scala naturae*? When for Russell, theology is situated at the top of all sciences,[7] this resembles very closely to this very traditional point of view. Nevertheless, the idea of a hierarchy of being in the world reflected by a hierarchy of sciences is also a key ingredient of almost any

5. Also Eörs Szathmáry identifies Polanyi's view of evolution 'as a progressive identification of the higher principles of life' (Polanyi 1968, 1308–12) with nothing less than the renowned idea of major transitions in evolution (Sigmund and Szathmáry 1998, 439).

6. For a more detailed discussion, see Losch (2017).

7. McGrath's idea, in contrast, is that theology is situated at the very fundament of the hierarchy of sciences (McGrath 2006).

critical realism (Losch 2009, 96–7). The vertical hierarchical arrangement is less a sign of dominance, however, than a sign of irreducibility (see also Cunningham, this volume), and another expression of the asymmetrical character of the relation between science and theology/religion. Peacocke's epistemic holism, as described by Russell,

> views the disciplines of the sciences and the humanities, including theology, as a series of emergent levels that reflect the increasing complexity of the phenomena they study. It involves two claims about these levels: 1) lower levels place *constraints* on upper levels (against 'two worlds' treatments that make them autonomous), but 2) upper levels are *emergent* and cannot be reduced entirely to lower levels (against 'epistemic reductionism'). Russell 2008, 7)

Here we have a similar concept as found in Polkinghorne's statement. That being said, without philosophical self-reflection this could lead to an unconscious physicalism. One must indeed take into account that all scientific data are theory-laden and, I would say, 'metaphysically stanced' as well. Finally, we should note that at least for Polanyi and Bechtel/Abrahamsen, religion is only *part* of the cultural phenomena, which constitute the highest level in the hierarchy;[8] it is *not* the queen of sciences.

Pathways of Interaction: Robert John Russell's Model of Creative Mutual Interaction (CMI)

Finally, we come to Robert John Russell's model of pathways of interaction, whose premises we now have illustrated. The idea behind this model was to think through the possibility of truly *mutual* interaction between theology and science, so not only a 'theology in light of science' as Barbour did, 'but *moving both ways between* theology and science' (Russell 2008, 21). A paradigmatic example for Russell is the case of Big Bang cosmology, and how Fred Hoyle's atheist belief inspired him to develop the steady state cosmology against the Christian appraisal of the Big Bang cosmology developed by a physicist and catholic priest, Georges Lemaître (Lambert, van Bibber and Ampleman 2015). The steady state theory is refuted today, but nevertheless, it was a scientific research programme inspired by personal (non)beliefs, by a *Weltanschauung* for Russell (Hoyle himself denied this, see Croswell 1996, 113–14).

Russell's method of CMI also presented diagrammatically in a pathway model (Russell 2008, 22; Peters, Russell and Welker 2002, 12):

8. Peacocke considers theology as a potential 'constitutional monarch' (Peacocke 1984, 37; see also Peacocke 1979, 367–71).

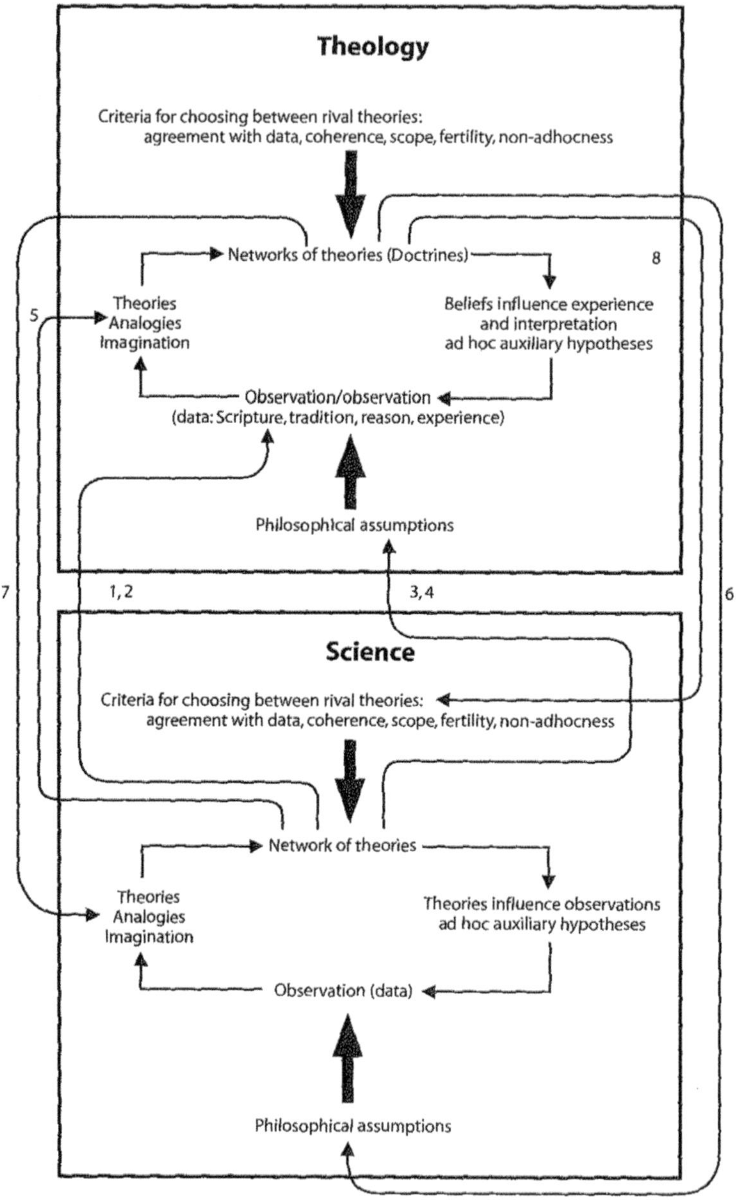

What interests us here first are not the great big arrows, but something no longer present from the diagrams developed previously. The direct downward-facing arrows *within the core rectangles* are missing, a sign of Russell's exclusive commitment to Lakatosian philosophy of science (Russell 2008, 16–20; see also Russell 2015), which is a sound critique of any assumption of direct falsifications.

Also, the additional entry 'ad hoc auxiliary hypotheses' in the right-hand loop underlines this observation.

What is more visible are two additional bold arrows, those below the data indicating 'philosophical assumptions', and those above the 'networks of theories' showing the criteria for choosing between rival theories. My comments on these additional arrows are the following: I am not sure whether the criteria in Barbour's original approach were not already symbolized by Barbour's (normal) downward arrows, which are now missing. If so, there would be no actual need to add another bold arrow on top of the structure. Also the other upward bold arrow at the bottom of each box, starting at 'philosophical assumptions' would perhaps be better interpreted as a part of the right-hand loop as in Barbour's original account. I would suggest, then, that Russell does not add anything significantly new in this regard, and indeed the many additional arrows just cause confusion. Robert John Russell himself answered these questions with the comment: The positioning of the bold arrows 'was to emphasize that the overall philosophical assumptions influence what data count (like astronomy vs. astrology)' (Russell, private communication). Here, he is of course right; although I maintain that one could integrate it into the original graphics.

New in Russell's diagram are also the eight thin long pathways indicating opportunities of exchange between the disciplines. As may be expected, five such paths lead upwards from science to theology. Following the idea of CMI, there are, however, also three paths leading downward.

Let us first have a look at the five pathways leading upwards, which is more in accordance with the overall structure of the diagram. This recalls the asymmetry between science and religion originally mentioned, which is expressed by this bottom-up movement.

In the first four, scientific theories can act as *data for theology* both in a direct sense (paths 1 und 2) and indirectly via philosophy (paths 3 and 4) (Russell 2008, 319). Also, scientific theories 'can function heuristically in the theological context of discovery by providing … inspiration' (path 5) (Russell 2008, 321). (The examples provided for these sorts of interactions are all from physics, as Russell is a physicist-theologian.)

1. Scientific theories 'can act directly as data which place constraints on theology'. This mirrors the basic structure of the diagram we developed. 'So, for example, a theological theory about divine action should not violate or ignore special relativity' (Russell 2008, 319).
2. Scientific theories 'can act directly as data either to be 'explained' by theology or as the basis for a theological constructive argument' (319). T = 0 (the beginning of time) in Big Bang cosmology could be explained theologically via creation ex nihilo. It should be noticed, however, that Russell himself is very careful not to claim a direct link here (see 40–2).
3. Scientific theories can act indirectly – after philosophical analysis – as 'data' in theology. 'An indeterministic interpretation of quantum mechanics can function within theology of divine action (special providence) by providing

a non-interventionist approach' (319) for God's action in the world. Russell himself is making use of this direction with his development of non-interventionist divine action (NIODA) (151–211).

4. Scientific theories can again act indirectly as 'data' in theology, 'when they are incorporated into a fully-articulated philosophy of nature' (321). An example would be Whitehead's philosophy of nature.
5. Also, scientific theories 'can function heuristically in the theological context of discovery by providing conceptual inspiration, experiential inspiration, moral inspiration, or aesthetic inspiration' (321). Big Bang cosmology may inspire a sense of God's immanence in nature, for instance.

These forms of interaction may all pass more or less undisputed. More troubling, however, is the question whether there can be any connection *downwards from theology towards science*. Wouldn't the historically developed atheistic methodology of science prohibit that?

Russell assumes at least three pathways from theology to science:

6. First, 'theology provides some of the philosophical assumptions which underline scientific methodology'. Russell's example here is the concept of a creation ex nihilo which 'played an important role in the rise of modern science by combining the Greek assumption of the rationality of the world with the theological assumption that the world is contingent' (321), thus giving birth to the empirical method and the use of mathematics in science.[9]
7. As the example of Big Bang and steady state cosmology demonstrates, 'theological theories can act as sources of inspiration … in the construction of new scientific theories' (321). Other examples provided (by Russell) are the theologies and philosophies that influenced the pioneers of quantum theory, 'including Vedanta for Schrödinger, Spinoza for Einstein,[10] Kierkegaard for Bohr' (321).
8. Finally, 'theological theories can lead to 'selection rules' within the criteria of theory choice in physics' (321). Russell's example here is the theological imagination of humankind being created in the image of God 'which includes libertarian free will and with it the possibility of enacting our choices bodily' (321). Thus (an indeterministic interpretation of) quantum mechanics may be preferred over classical mechanics. This is a very prevalent way of thinking in the field of science and religion, which, however, does not mean that it is unproblematic.

So, while the naturalism essential to science prohibits any reception of theology within the *method* of science, one has to admit a personal component in the structure of science, so there can be an indirect influence by theological or metaphysical concepts on the *persons* doing science.

9. See John Haught's approach of 'confirmation' in Haught (1995).
10. For a critique of the claim of Spinoza's influence on Einstein, see Mühling (2011).

A more prominent example than the aforementioned Fred Hoyle would probably be the case of the inventor of the Big Bang theory, physicist and priest Georges Lemaître. Dominique Lambert observes:

> Probably the theme of light as the primordial stuff of the universe has played a role in Lemaître's imagination. The biblical context is not used directly in Lemaître's papers of course. But it is interesting to note that intuitions or images coming from his symbolic exegesis contribute to provide insights in a field he explored and where he did not know anything.

Lambert continues by analysing the systematic role of theology in this context:

> Religious background plays the role here of providing intuition. Of course, when the theory was established, all these extra-scientific insights disappear with the context of discovery, leaving the place only to logical and empirical elements belonging to the scientific context of justification. Science practice, however, is not only built on pure rational and logical foundations. When entering in unknown areas scientists need to be helped by some guides, some regulating ideas. … Imagination, aesthetic considerations, philosophical assumptions and even religious (or sometimes anti-religious) prejudices can play some role providing such transitory guides. At this point needs to be mentioned, however, that such 'philosophical' or 'religious' input has not to be confused with what we could call the real metaphysical structure or content of science. (Lambert 2017)

Lemaître was also trained in philosophy. Maybe even more important than his religious beliefs, this training allowed him to avoid an often-prevalent conflation and prepared him to *distinguish* between a natural beginning of the universe and its metaphysical start. 'Thomistic philosophy gives him a tool to avoid an epistemological confusion and to feel free to explore cosmological models with initial singularity' (Lambert 2017).

Concluding Questions

Although his examples would have to be checked by historians of science, Russell has to be congratulated for exploring the idea of CMI between science and religion, and stressing the 'downward' move from theology to science. The importance of the *role of philosophy* in a substantial theory of modern physics would maybe demand a more intense treatment of philosophy in portraying the relation between science and theology than Russell already provides. Philosophy, whether in the form of popularized metaphysics or of knowledge acquired through traditional training, always plays an important role in relating science and theology. The whole theory developed by Russell and others is itself a form of philosophy, a *metaphysical theory*. Philosophy permeates the whole diagram so to say. That's why it is important to have philosophers as part of any science and religion discourse.

Barbour, Peacocke, Polkinghorne and Russell all chose critical realism as their basis for their philosophy of science. I am convinced this philosophical stance has in itself some substantial shortcomings. I have criticized it elsewhere for addressing 'the similarities between the methods of science and religion, but it does not take into account the differences between the two' (Losch 2005, 280). Against Barbour's spectrum thesis of only gradual – that is quantitative – differences, I advocate *qualitative* or even *categorical* differences between natural science and theology. It is my critique that in-between the two disciplines, social and human sciences also play an important role. While every science can be hermeneutically conceived, social and human sciences operate with sort of a double hermeneutics. 'Social scientists do not only have to regard their own subjectivity, they also need empathically to understand their objects of study as a necessary precondition for exploration and explanation; in addition to the interpretative framework imposed by every researcher, the object of research here is itself symbolically structured' (Losch 2005, 282, referencing Clayton 1989, 101).

Additionally, in theology its 'object' demands a treatment suitable for its structure as being actually a subject, a personal counterpart (Losch 2016). Here, I assume it better not to speak of a verification (or Popperian falsification) of knowledge any more, but to use with Michael Polanyi the term 'validation'. Both verification and validation 'are an acknowledgement of a commitment: they claim the presence of something real and external to the speaker. The structure of commitment remains unchanged, but its depth becomes greater; when we pass from verification to validation, we rely increasingly on internal rather than external evidence' (Losch 2005, 280, referencing Polanyi 1998, 321). I tried to integrate this critique elsewhere into a proposal of a *constructive-critical realism* as a more adequate theory of relating science and theology, including the humanities (Losch 2005, 2018). I am aware that I am philosophizing by doing so.

References

Barbour, Ian G., *Christianity and the Scientist.* The Haddam House Series on the Christian in His vocation (New York: Association Press, 1960).

Barbour, Ian G., *Issues in Science and Religion* (Englewood Cliffs, NJ: Prentice-Hall, 1966).

Barbour, Ian G., *Religion and Science: Historical and Contemporary Issues.* 1st HarperCollins rev. ed. (San Francisco: HarperSanFrancisco, 1997).

Barbour, Ian G., 'Taking Science Seriously without Scientism: A Response to Taede Smedes', *Zygon*® 43, no. 1 (2008): 259–69. doi:10.1111/j.1467-9744.2008.00911.x.

Barth, Karl, *The Doctrine of Creation: (Church Dogmatics, Volume III, 1).* Church Dogmatics / by Karl Barth 3.1 (Edinburgh: T. & T. Clark, 1958).

Bechtel, William, and Adele A. Abrahamsen, *Connectionism and the Mind: An Introduction to Parallel Processing in Networks* (Cambridge, MA, US: B. Blackwell, 1991).

Clayton, Philip, *Explanation from Physics to Theology: An Essay in Rationality and Religion* (New Haven, CT: Yale University Press, 1989).

Clicqué, Guy M., *Differenz und Parallelität: Zum Verständnis des Zusammenhangs von Theologie und Naturwissenschaft am Beispiel der Überlegungen Günter Howes*. Untersuchungen zum christlichen Glauben in einer säkularen Welt Bd. 1 (Frankfurt am Main, Berlin, Bern, Bruxelles, New York, Oxford, Wien: Lang, 2001). Nürnberg, Univ., Diss--Erlangen, 2000.

Croswell, Ken, *The Alchemy of the Heavens* (Oxford: Oxford University Press, 1996).

Dinter, Astrid, *Vom Glauben eines Physikers: John Polkinghornes Beitrag zum Dialog zwischen Theologie und Naturwissenschaften* (Mainz: Matthias-Grünewald-Verl., 1999). Univ., Diss.--Marburg, 1998.

Harrison, Peter, *The Territories of Science and Religion* (Chicago and London: The University of Chicago Press, 2015).

Haught, John F., *Science and Religion: From Conflict to Conservation* (New York [u.a.]: Paulist, 1995).

Henrich, Jörn, *Die Fixierung des modernen Wissenschaftsideals durch Laplace* (Berlin: Akademie Verlag, 2010).

Lambert, Dominique, 'Was the Big Bang Theory Born Out of Belief? Lemaître's Primeval Atom Hypothesis'. Accessed 21 August 2017. http://www.theologie-naturwissenschaften.de/startseite/leitartikelarchiv/big-bang.html

Lambert, Dominique, Karl van Bibber and Luc Ampleman, *The Atom of the Universe: The Life and Work of Georges Lemaitre* (Portland: Copernicus Center Press, 2015). http://gbv.eblib.com/patron/FullRecord.aspx?p=2095569

Losch, Andreas, 'Our World is More than Physics: A Constructive – Critical Comment on the Current Science and Theology Debate', *Theology and Science* 3, no. 3 (2005): 275–90. doi:10.1080/14746700500317271.

Losch, Andreas, 'On the Origins of Critical Realism', *Theology and Science* 7, no. 1 (2009). 85–106. doi:10.1080/14746700802617105.

Losch, Andreas, 'Critical Realism – A Sustainable Bridge Between Science and Religion?' *Theology and Science* 8, no. 4 (2010): 393–416. doi:10.1080/14746700.2010.517638.

Losch, Andreas, *Jenseits der Konflikte: Eine konstruktiv-kritische Auseinandersetzung von Theologie und Naturwissenschaft*. Forschungen zur systematischen und ökumenischen Theologie 133 (Göttingen [u.a.]: Vandenhoeck & Ruprecht, 2011).

Losch, Andreas, 'Wissenschaftliche und religiöse Welterfassung: Ein Kommentar zu Martin Bubers Perspektive auf ein spannungsreiches Thema', *Theologische Zeitschrift* 70, no. 2 (2014): 142–61.

Losch, Andreas, 'Kant's Wager', *International Journal of Astrobiology* 15, no. 4 (2016): 261–70.

Losch, Andreas (ed.), *What Is Life? On Earth and Beyond* (New York: Cambridge University Press, 2017).

Losch, Andreas, 'Appreciating Faith and Culture in an Age of Scientific Reasoning. On Constructive-Critical Realism', *Theology Today* 75, no. 2 (2018): 154–66.

McGrath, Alister E., *A Scientific Theology: Reality* (London: T & T Clark, 2006).

Mühling, Markus, *Einstein und die Religion. Das Wechselverhältnis zwischen religiös-weltanschaulichen Gehalten und naturwissenschaftlicher Theoriebildung Albert Einsteins in seiner Entwicklung*. 1. Aufl. Religion, Theologie und Naturwissenschaft Religion, theology, and natural science Bd. 23. s.l.: Vandenhoeck Ruprecht, 2011. http://site.ebrary.com/lib/alltitles/docDetail.action?docID=10519359

Peacocke, Arthur Robert, *Creation and the World of Science: The Bampton Lectures, 1978*. Bampton lectures 1978 (Oxford: Clarendon Press, 1979).

Peacocke, Arthur Robert, *Intimations of Reality: Critical Realism in Science and Religion.* The Mendenhall lectures 1983 (Notre Dame, IN: Published for DePauw University), Greencastle, Ind., by University of Notre Dame Press, 1984.

Peacocke, Arthur Robert, *Theology for a Scientific Age: Being and Becoming–Natural, Divine, and Human.* 1st Fortress Press ed. Theology and the Sciences (Minneapolis: Fortress Press, 1993).

Peters, Ted, *Science and Theology: The New Consonance* (Boulder, CO: Westview Press, 1998).

Peters, Ted, Robert J. Russell and Michael Welker, *Resurrection: Theological and Scientific Assessments* (Grand Rapids, MI: W.B. Eerdmans Pub, 2002).

Polanyi, Michael, *Science, Faith and Society.* Riddell Memorial Lectures 18 (London: Cumberlege, 1946).

Polanyi, Michael, 'Life's Irreducible Structure: Live Mechanisms and Information in DNA Are Boundary Conditions with a Sequence of Boundaries Above Them', *Science* 160, no. 3834 (1968): 1308–12. doi:10.1126/science.160.3834.1308.

Polanyi, Michael, *Personal Knowledge: Towards a Post-Critical Philosophy* (London: Routledge, 1998).

Polkinghorne, John, *Science and Theology: An Introduction* (London, Minneapolis, MI: SPCK; Fortress Press, 1998).

Polkinghorne, John, 'Physics and Metaphysics in a Trinitarian Perspective', *Theology and Science* 1, no. 1 (2003): 33–49. doi:10.1080/14746700309645.

Russell, Robert J. (ed.), *Fifty Years in Science and Religion: Ian G. Barbour and His Legacy.* With the assistance of Ian G. Barbour. Ashgate science and religion series (Aldershot: Ashgate, 2004).

Russell, Robert J., *Cosmology: From Alpha to Omega the Creative Mutual Interaction of Theology and Science.* Theology and the sciences (Minneapolis: Fortress Press, 2008).

Russell, Robert J., 'The Crucial Importance of Nancey Murphy's Deployment of Lakatos's Methodology for Theology and Science', in *Practicing to Aim at Truth: Theological Engagements in Honor of Nancey Murphy*, ed. Ryan A. Newson (Eugene: Wipf and Stock Publishers, 2015), 13–25.

Sigmund, K., and E. Szathmáry, 'Biomathematics: Merging Lines and Emerging Levels', *Nature* 392, no. 6675 (1998): 439. doi:10.1038/33020.

van Orman Quine, Willard, *From a Logical Point of View: 9 Logico-philosophical Essays*, 2nd edn, rev., reprint. Harper torchbooks. The @science library 566 (New York: Harper & Row, 1963).

Whitehead, Alfred North, *Religion in the Making: Lowell Lectures, 1926* (New York: The Macmillan Company, 1926).

Chapter 4

WHO'S AFRAID OF REDUCTIONISM'S WOLF? THE RETURN OF SCIENTIA

Conor Cunningham
University of Nottingham

Science is an answer to a question that precedes it

(Sergei Bulgakov)[1]

'Knowledge also is surely one, but each part of it that commands a certain field is marked off and given a special name proper to itself. Hence language recognizes many arts and many forms of knowledge.'

(Plato)[2]

Introduction

This chapter argues several mutually reinforcing points, many of which lead to the conclusion that both scientism and reductionism are false. The temptation to indulge such erroneous philosophical positions arises from our bewitchment by a picture of a hierarchical view of the sciences, with physics being the master science. Physics is that which is really real (ontas onta), to echo Plato, while all others pale into various degrees of insignificance. This is the idea of the layer cake. Such a highly misleading picture itself rests on a fiction; that of a base. From the pre-Socratics to now, this perennial temptation takes on many forms, whether it be Thales' water, Democritus' atom, Dawkins' DNA, the Brain, or physics and the 'particulate'. In short, I argue that there is no base (except that of *methexis*), no fundamental level that would afford the fundamentalism of a grand science, concomitantly accommodating reduction to that science's base element or currency, so to speak.

1. Bulgakov 2000, 170.

2. *Sophist*, 257c. Aristotle in *On the Heavens* echoes this, as does Aquinas in his *Commentary on Boethius' De Trinitate*, wherein he, like his forebears, argues for an alethic monism, but one that is intrinsically pluralist.

The above critique is bolstered by two central arguments. Philosophy tends to operate in the opposite direction from physics, especially when it comes to reduction and theory construction. Consequently, many philosophical problems, such as supervenience or causal inheritance, are wrongheaded, if not old fashioned. Furthermore, the temptation for reductionism should be resisted; yes, but reduction itself should not be feared: While emergence is most certainly prevalent, but it is not *always* preferable, and can at times act as nothing but a placeholder. Likewise, bottom-up causation versus top-down causation are not to be set in strict opposition. The relation between the micro and macro is not asymmetrical, as there is no base on which they sit. At times the macro is in charge, while at other times the micro is (by micro we mean the primitive subvenient material base, for example, the particulate). The relation can be symmetrical, no doubt, as the micro is often beholden to the macro, yet the macro can also be beholden to the micro, indeed it can even be a 'part' of the micro (see below). The whole may well indeed be more than its parts, being so because it is different than its parts, but that is only very significant for the advocate of reductionism. Lastly, mechanism is not necessarily in opposition to the non-mechanistic, or reductionist and anti-reductionists, in short, we need both.[3] The chapter's conclusion is that scientific ontology is plural, not to mention its epistemologies, all of which are beholden to an alethic monism, metaphysically speaking. All *scientia* involves, without doubt, an apophatic and a cataphatic moment: we posit a primitive term, doing so with trust; but in so doing we know we fail to capture.

Divorce Cake – Scientism and Reductionism

The origin of the division of the sciences into isolated fiefdoms which are self-governed is a very modern idea, which can be traced back to the juridical model of the sciences that emerged in the thought of Bartolo da Saasoferrato in the fourteenth century, and was later formulated in more explicit terms by Jean Bodin in 1576, wherein separate sciences lived under the organizing principle that *superiorem non recognoscens* – each science does not recognize any superior authority. This way lay autonomy, a sort of aseity. It was presumed that the sciences consisted in a layer cake, to borrow Putnam and Oppenheim's metaphor from the 1950s. Under the sway of this image reductionism and its sibling scientism were

3. For example, in relation to cancer we adopt a cell-based approach which involves reduction, namely, somatic mutation theory. At the same time, we must adopt a tissue-based approach which is more about carcinogenesis, namely, tissue organization field theory (see Bertolaso 2016). Similarly, in nuclear physics we have two models, namely the nuclear shell model and the liquid drop model. The latter treats the nucleus as an incompressible drop of nuclear fluid, and this does not afford mechanistic explanation, while the former does insofar as the nucleus is approached in terms of energy levels. Both are to be employed.

born.[4] Science is the only begetter of truth (first step), but in truth all sciences were to be reduced to one master science, physics (second step). Yet these two steps leave science vulnerable. As van Fraassen asks, 'Can we divide our language into a theoretical and non-theoretical part? … All our language is thoroughly theory-infected' (van Fraassen 1980, 12). The precarious nature of all theorizing, including that of the sciences, comes immediately to our attention when theories change, especially in the case of *radical theory change* (we need to only think about the move from Newtonian physics to Einstein's special and then general relativity (GR), or the shift from classical to quantum mechanics), which is evidence of 'the transience of our best accepted theories' (Sklar 2010, 1122). Hence we can speak of *theories emeritus*. What was taken to be obvious turns out not to have been the case. This is sometimes referred to as 'meta-pessimistic induction': things we took as settled in science changed, therefore science never rests on a permanent ground. In short, we must *not* base our philosophies on what, for example, current physics says (or indeed any science), extrapolating philosophical conclusions from such a ground without knowing what future physics will tell us (Hempel's dilemma), *and* we must not fall into the trap of thinking that qualitative, *radical theoretical change* is over, as so many under the sway of Newtonian physics once thought. In this way the claim for a 'base' to the layer cake is truly undermined: Science is, to paraphrase Poincaré, ruins accumulated upon ruins.

In 1949 Ernst Nagel published his understanding of what reduction entailed, made famous later (though unchanged) in his 1967 book (see Nagel 1949; 1967). According to Nagel, reduction was a matter of intertheoretic explanation, which is to say, the reduced theory is explained by the reducing theory: explanation being understood in deductive-nomological terms (and its extensional equivalence), which is in the end a degenerate form of statistical explanation (see Wimsatt 1979). Reductions were either homogeneous, such as the reduction of Galilean laws of falling bodies to Newtonian mechanics (Nagel 1967). Or they were heterogenous, for example, temperature being reduced to statistical mechanics. Generally, Nagel argued for two requirements for reduction: connectability and derivability. In truth, as Sklar pointed out in 1967 (and Nickles soon after confirmed his findings) this is naive from a historical point of view. The number of actual cases in the history of science where a genuine homogeneous reduction takes place are very few and far between, and as for heterogenous, they are just as problematic, most of the time. Indeed, below we will address the example just provided, demonstrating its failure (Sklar 1967; Nickles). Important here, however, are the philosophical implications of such reductionism. Scientism, but not *scientia*, faces the following charge: All sciences *Vicissim omnes ab omnibus repelluntur* ('all is in turn refuted by all') to quote Jean Bodin from his *Colloquium Heptaplomores* (1593). The sciences are very high forms of art, no doubt, art that necessarily involves interpretation. But 'it is crucial to observe that the demand for interpretation arises within theoretical

4. The term 'reductionism' first appears in Garnett (1942). Scientism appears to have been first used by Friedrich Hayek.

science. It isn't something imposed on the scientist by some aprioristic philosopher' (Sklar 2010, 1123). The Quine–Duhem thesis that theories are underdetermined by the evidence should be the bedtime story of every fundamentalist. After all, we've more than thirty theories of the atom. It should be recalled that soon after Paul Dirac published his relativistic wave equations for the electron in 1928, Max Born declared that 'physics, as we know it, will be over in six months'. Blushes all round, back then at least. But not in 1980 for Hawking who this time proposed that Gauge Field Programme would be the end of the story (see Hawking). Today we have several candidates: String theory; M-Theory, the ADS/CFT correspondence (also called the 'holographic conjecture').

In physics the rise of the standard model (SM – electroweak theory and quantum chromodynamics) in the 1970s and 1980s was interpreted as a major success for reductionism, that is, the move to a 'final theory', forgetting that even then gravity was missing (quantum gravity still eludes us, of course) and many parameters helping to construct the model were arbitrary. After many developments in *quantum field theory* (QFT), and the employment of renormalization group theory, our understanding of the SM has changed radically. It is now construed as an *effective field theory* (EFT). An effective theory (ET) is that which captures what is relevant physically in, or at, a given domain, doing so effectively – it works, and it does so by ignoring all else (see Georgi 1995, 88). This results in a natural pluralism. Supporting this, Cao offers a vision, 'I would like to define a fundamental theory not as a theory from which we can derive all other theories, but rather, as a theory that cannot be derived from any other theory. In this sense QFT definitely is a fundamental theory. In the same sense the general theory of relativity is also a fundamental theory' (Cao 2003, 28). In physics sometimes the fundamental theory is characterized in terms of the tiniest length-scale or highest energy scale; by contrast, the fundamental theory is one that is coherent with all other accepted physical principles. The problem here being that GR fails on both counts, while QFT fails to meet the second criterion. Why? Because it does not incorporate the dynamic understanding of space-time that GR accepts (Cao 2003, 29). More crucially, physics does not reduce to physics (see van Brakel 2014, 33). We need physicists, after all. This is a version of the Quine–Putnam indispensability argument – the physicist is here indispensable, as are many levels of analysis, energy levels, and so on. Any such physicists leave their beds each morning for a reason, a good one, we wager, and head off to the lab, lecture hall, or to their desk, pursuing that which is deemed good, beautiful and true. As Aristotle tells us, 'For the good and the beautiful are the beginning both of the knowledge and of the movement of many things' (*Metaphysics*, 1013a 22–3).

Reductionism and Hierarchies

From the above we can begin to glean that *reductionism is false* – historically, philosophically, and scientifically. It is, moreover, a metaphysical claim. Indeed, we should wonder how reductionism is to *utter* data at all, given its own terms, for all components of said utterance would fall into the disarray of an insipid atomism. The sense of our concepts lapses into nominalism. Aquinas argues as much while expounding Aristotle (*Metaphysics*, VII.17.1673–4). Reductionism as a purveyor of ontological fundamentalism or microphysical fundamentalism with its requisite base element looks something like this:

1. The hierarchy thesis: The universe is stratified into levels.
2. The fundamentality thesis: There is a bottom level, which is fundamental.
3. The primacy thesis: Entities on the fundamental level are primarily real and the rest are at best derivative, if they are real at all (see Schaffer 2003, 498).

Attendant to the above are four interpretative frameworks for the idea of 'level': *mereological*, part to whole,[5] wherein the whole is composed by its parts; *supervenience*, which invokes asymmetrical relations of dependency, the higher being wholly indebted to the lower; *realization*, here, all functions are realized in lower level properties; lastly, *nomological*, this simply means there are *one-way* bridge laws between levels, in which the higher is reducible to the lower (Schaffer 2003, 498–517). Here we have our layer cake:

> The hierarchy of levels is a hierarchy of sets of entities corresponding to the universes of discourse for different branches of science. This hierarchy is arranged so that the branch of science dealing with level *i* is a 'potential micro-reducer' of the branch of science dealing with the next highest level *i* +1 in that the entities from the higher level can be decomposed into entities at the lower level. It is further stipulated that any whole that's exhaustively decomposable into parts belonging to level *i* also belongs to *i*; thus, the population at each level includes all entities at higher levels. The hierarchy of layers is therefore ontologically conservative in the sense that moving up in the ordering does not add any ingredients to the world: all entities are already contained in the bottom layer, the level described by the most basic theories in physics. (McGivern and Rueger 2010, 381)

The temptation of hierarchy provides a sort of artificial neatness. What it reveals, quite clearly, is philosophical prejudice. It does so because the order of discovery is the reverse of what it should be: we discover hierarchy before we look. Interestingly, when physicists speak about levels it's more a question of the stratification of reality into processes, scales, behaviours and so on, rather than that of hierarchy as such (see McGivern and Rueger 2010, 381). Nickles' most telling observation is that there are two forms of reduction (I would argue there are many more, such as cultural). Let us call this *reduction in two flavours:* that of physics and philosophy, which portray an annoying terminological orthogonality (see Berry 1994, 598). First, in physics a typically newer and more refined theory is said to 'reduce to' an older and coarser theory: a finer theory such as relativity does in a sense 'reduce to'

5. 'Mereology' was coined by Léniewski (1886–1939).

Newtonian physics under the 'conditions' that supported the Newtonian theory. Relativity is thus said to 'encompass' Newtonian theory. Philosophy, meanwhile, tends to have a higher, less general theory, which is older, 'reduce' to a lower, more general, and more recent theory. Against the direction assumed by philosophy, consider, for instance, that certain features of rainbows can be fully understood only through asymptotic methods: in effect, there are universal features that emerge in the asymptotic domain as the wave theory approaches the ray theory in the limit as the wavelength of light approaches zero. These features inhabit, so to speak, an asymptotic borderland between theories (see Batterman 2002, 9–22; Chibbaro, Rondoni and Vulpani 2014, 31–2). The phenomena inhabiting this borderline are not explainable in either purely wave-theoretic or ray-theoretic terms. A full explanation of what we observe in the rainbow cannot be given without reference to the asymptotic domain between wave and ray theories. The relevant phenomena are simply not reducible to either theory: to neither the more 'fundamental' wave theory nor the 'coarser' ray theory. They cannot be predicted from the properties of the more fundamental theory, even though they are in a sense asymptotically contained in that theory. This should give pause to much philosophical ambition of recent times.

You Can't Have Your Ingredients Without Your Cake – Macro Phenomena and Renormalization

Another challenge to hierarchical reduction comes in the form of the relationship between the macro and the micro level. Batterman's approach to the question of multiple realization (made famous by Jerry Fodor) is promising, insofar as he endeavours to answer the perennial conundrum of how heterogenous systems at the micro-scale are homogenous in terms of their behaviour at the macro level. This, quite correctly, is a major challenge to reductionism. He does by invoking the phenomenon (an emergent one, as we shall see) of renormalization (especially, renormalization groups – RGs) (see Wilson). To repeat, how are there stable robust behaviours at the macro when there is no micro base for such a phenomenon, and micro realizers seem to be largely ignored? An example of this would be how theories of continuum scale physics (e.g. continuum mechanics, thermodynamics, or hydrodynamics, etc.) do what they do, and do it so well without appeal to the micro (Batterman 2017, 6). Orthodox thermodynamics is phenomenological, insofar as it treats a system, such as a gas in a box, as a continuous blob of stuff, thus when it describes (and explains) the observable behaviour of various systems like gases, it remains agnostic, indeed aloof about the internal constitution of those systems. Conversely, kinetic theory (and statistical mechanics) explains behaviour of a gas in terms of fundamental principles, but any success rests on forfeiting the encounter of the very phenomenon to be explained.

In dealing with macro phenomena, especially exhibited behaviour, renormalization steps in. For example, as Jackiw notes, 'ultraviolet infinities appear to be intrinsic to quantum field theory, and no physical consideration can circumvent them; unlike the infrared [low energy] divergences, ultraviolet ones cannot be excused away. But they can be "renormalized".' (Jackiw 1999, 149–50). By appealing to the terminology of physics rather than that of philosophy new insights are afforded. The main terminological shift (though with very similar meaning) is from multiple realization to that of 'universality'. Berry puts it this way, universality is 'the slightly pretentious way in which physicists denote the identical behaviour in different systems. The most familiar example of universality from physics involves thermodynamics near critical points' (Berry 1987, 185). Put simply, different systems that consist in wholly different micro constituents do in certain circumstances act as if they were the same physical system.[6] For example, vastly different liquids (e.g. neon (Ne) and methane (CH_4)) converge on the same behaviour under particular circumstances, rendering reductionism in a bit of a quandary. Another way to put this is that one can 'change the "essential molecular features" of, say, methane into those of neon, without affecting the upper-scale behaviour .... This reflects a kind of *stability under changes* of the very nature of the systems. The renormalization group makes the metaphor of "morphing" one system into another mathematically precise' (Batterman 2017, 8). By way of an averaging rule the interactions between molecules are reigned in because of its enormity – averaging out allows for this as it begins the process of decreasing freedom otherwise the freedom is so vast, indeed infinite, that it lies beyond calculation in any meaningful sense. These averages then act or are treated as something 'real' or, better, as new molecular components, which makes sense because of the self-similarity displayed by systems near the crucial point (crudely, they sort of look like something, they manifest *form*, so to speak) (Batterman 2017, 9). In a sense one coarse grains one's analysis (like when we half close our eyelids to see better), screening-off that which is no longer relevant (Batterman 2017, 9). Moreover, the underlying high-energy physics is multiply realized, as it were, to the point of near indifference. In other words, any macro description that such a procedure provides involves a strong compression of data, a strong selectivity (see Morrison 2014, 1145). As we know, this whole procedure is one of art: 'There are no recipes for how to reconstruct an RG for a given universality class.' This being the case the reductionist is once again in trouble. 'Reductionists can never concede that coming up with the right blocking procedure is an art.' Consequently, renormalization is a liberator of the macro. This emancipation involves the lifting of the macro to a new realm or order, as Thalos calls it, but without jettisoning crucial relations with other realms; it is a marriage of equals (Thalos 2013, 239).

6. An analogy is to be found in evolutionary biology, namely the phenomenon of convergence (or homoplasy). There, unconnected creatures 'solve' evolutionary problems in the same way, suggesting that the biosphere is constrained: here the creatures are significantly diverse in the physical makeup (genotype, phenotype, environment, etc.), yet they end up at the same party, so to speak. A good example is that of the camera eye, which is found in humans and in octopus also.

Priority of the Pie – Autonomous Ontological Domains

In examining the relationship between the macro and the micro level we need to pay attention to the ontological status given to each (or indeed to the whole). Schaffer uses QFT to offer an interesting critique of microphysical fundamentalism. He posits what might be termed the thesis of 'infinite descent'. The purpose of which is to argue that an analysis that seeks a base element could always go further, and history seems to chime with his contention. According to Schaffer, 'Infinite descent yields an egalitarian metaphysics which dignifies and empowers the whole of nature. Treat infinite descent as a working hypothesis, and since all entities turn out to be composite, supervenient, realizes, and governed, it emerges that these attributes cannot be barriers to full citizenship in the republic of being' (Schaffer 2003, 51). The point being that if the world is infinitely decomposable then if there is any halting or drawing the line at a certain level it is question begging. If we do select a level as a dependence base, then we transform all that which might be below it into something nonphysical, and surely the physicalist does not wish to have such phenomena in its ontology. Schaffer baulks at the usual presentation of monism as the thesis that there is just One because according to him we are reading this all wrong. It is not that there is just One, but rather that the One is *prior* to its parts (priority monism). Here Schaffer is following the lead of Proclus, consciously so (see Proclus 1979, 79).

Put another way, *emergence* is metaphysically possible, *submergence* is not. If it were possible then the intrinsic properties of the proper parts, and the fundamental relations between these parts, must fail to supervene on the intrinsic properties of the whole, in other words, they stand alone untouched. For Schaffer, this is impossible because 'an intrinsic property of the proper pars ipso facto correlates to and intrinsic property of the whole, namely, the property of having-a-part-with-such-and-such-intrinsic property, and relations between the parts also correlates with an intrinsic property of the whole, namely, the property of having-parts-thus-and-so-related. Fix the whole, and all of its parts are fixed' (Schaffer 2010, 56). The point being that unlike the pluralist who is vulnerable to emergence the monist can provide an inventory of the world (Ciauncia-Garrouty 2013, 568).

Schaffer also appeals to the *asymmetry of existence*, which simply means that there must be a basic whole, but there need not be basic or ultimate parts. Schaffer's thesis of infinite descent is empirically supported, hence quantum entanglement, and it is, in addition, an empirically open scenario (Ciaunica-Garrouty 2013, 568). Interestingly Schaffer, too, invokes renormalization in QFT wherein, and following Georgi, EFTs 'might form an infinite tower which goes down to arbitrary short distances in a kind of infinite regression … just a series of layers without end' (Georgi 1989, 456). Consequently, the world is approached as layered into quasi-autonomous domains (or realms, as Sarkar prefers) (Sarkar 1998, 193, fn. 22) and each domain has its own ontology and its own fundamental laws (see Cao 1997). Laughlin and Pines baptize this insulation from micro reductions 'protectorates' (see Laughlin and Pines 2000, 28–31). Within such protectorates, scientific understanding need not rely on ever-decreasing scales of investigation. The emergent phenomena gain as much ontological status as the elementals, so-called. For Laughlin and Pines something as seemingly simple as the existence of sound in a solid is emergent, which is to say, we do not appeal to microstructures: 'It is rather obvious that one does not need to prove the existence of sound in a solid, for it follows from the existence of elastic moduli at long length scales, which in turn follows from the spontaneous breaking of translational and rotational symmetry characteristic of the crystalline state. Conversely, one therefore learns little about the atomic structure of a crystalline solid by measuring its acoustics' (Laughlin and Pines 2000, 29). As Nelson points out,

> The modern theory of critical phenomena has interesting implications for our understanding of what constitutes 'fundamental' physics. For many important problems, a fundamental understanding of the physics involved does not necessarily lie in the science of the smallest available time or length scale. The extreme insensitivity of the hydrodynamics of fluids to the precise physics at high frequencies and short distances is highlighted when we remember that the Navier-Stokes equations were derived in the early nineteenth century, at a time when even the discrete atomistic nature of matter was in doubt. (Nelson 2002, 3)

Importantly, Bitbol makes the point that 'the "never ending tower" of autonomous domains in Quantum Field Theory indeed concerns domains of study, domains of concepts, but not domains of being' (Bitbol 2007, 302). In short, we must not reify (see below), and this stands for all approaches in *scientia*.

> If the message of quantum physics is taken seriously, the critique of reification concerns not only the high-level properties but also the low-level properties; not only the emerging properties, but also the properties of the so-called basic constituents of the world. The reductionists eventually lose the game, because their so-called 'reduction basis' is as firm as quicksand, and because it proves quite easy in this case to put the emergent behaviour on exactly the same footing as the so called 'elementary' entities and laws. (Bitbol 2007, 295)

Weighing the Scales

While we have here established something of a democracy of scale, Thalos goes yet further and argues that there is no true scale to which activity is confined. Of course this has enormous ramifications for reduction. But also, according to Thalos, it also offers a challenge to emergence, which she thinks concedes too much to reductionism, and thereby corrupts their own insights: it carries the shadow of that which it rejects. This is, as Thalos argues, reductionism in sheep's clothing (Thalos 2013, 42). The universe is scale free, therefore, without hierarchy, it being more of a case by case matter, as it were: a matter of context, but not contextualism. Most importantly, locality is denied: it is not a universal feature of the universe, which is to say the move to reduce to 'somewhere' is gibberish (entanglement would surely support this contention) (Thalos 2013, 56). Not only is the base level unnecessary, there cannot be any fundamental level. Gone is the very notion of a base element, a fundamental level, a hierarchy, and by implication, the division of the sciences into special and otherwise, for there is no 'master science' (Thalos 2013, 7). Anderson concurs, even though he accepts reductionism, which for him means the ability to reduce everything to simple fundamental laws, he rejects what seems to be asserted concomitantly, namely, the constructionist hypothesis, which simply asserts that we can reconstruct the universe from those laws, which is impossible. In this sense reductionism becomes domesticated (see Anderson 1972, 393–4). And domestication is a correct term, for as Anderson points out, the more reductionism succeeds the less is achieved (see below). Weinberg in his fight with low-energy physics over the funding of the superconducting super collider argued that 'particle physics is in some sense more fundamental than other areas of physics' (Weinberg 1987, 434).[7] Such an understanding reflects Weisskopf's distinction between extensive and intensive research (Weisskopf 1965, 24). Anderson's riposte is most revealing: 'The more the elementary particle physicists tell us about the nature of the fundamental laws, the less relevance they seem to have to the very real problems of the rest of science' (Anderson 1972, 393). Condensed matter physics has cut into the nomological hegemony of high-energy physics (Humphreys 2016, 6).

Eddington famously challenged any attempt to reduce to either macro or micro, by asking us to choose between two accounts of a table: the micro physical, or the common-sense macro description. Are the macro properties of the table such as rigidity and solidity reducible to the micro-properties, do they supervene? But even this question is inadequate. There is no common denominator that provides the necessary structure for such a question to be asked. It is only by first accepting, which is just presuming, the hegemony of micro-physicalism or generative atomistic physicalism that any dilemma gains a semblance of intelligibility (Humphreys 2016, 16). The macro, emergent properties are global in scope, and as such can only be approached from a global perspective, for once again the micro cannot understand about what you are speaking. It must be remembered that different energy levels quite literally display remarkably different physics. We witness this insofar as QFT, wherein EFT is the correct way of describing the relevant physics in particular limited energy domains, is a matter of context and approximation, and in so being it is more real. This Cao points out is 'a pluralism in theoretical ontology, an anti-foundationalism in epistemology and an antireductionism in methodology' (Cao and Schweber 1993, 69).

Outside mathematics one would be hard-pressed to find an example of true intertheoretical reduction (see Howard 2007, 145). The example of the reduction of macroscopic thermodynamics to classical statistical mechanics is a telling demonstration of the impossibility of reducing macro to micro, or vice versa

7. For the politicization of science see O'McGarity and Wenger (2010).

(e.g. the reduction of the ideal gas law PV = nRT to statistical Mechanics). It does not work. 'Foremost among the thermodynamic laws that must be derivable from statistical mechanical postulates is the second law, which asserts the exceptionless evolution of closed non-equilibrium systems from states of lower to states of higher entropy. Providing a statistical mechanical grounding of the second law was Boltzmann's paramount aim in the latter part of the nineteenth century.' Howard asks, did he succeed? No, because what Boltzmann derived was a statistical simulacrum of the second law of thermodynamics. All this simulacrum can tell us is that closed non-equilibrium systems are at best highly likely to evolve from states of lower to states of higher entropy (Howard 2007, 146). More importantly, this statistical understanding is not deduced from first principles, or at least only when such principles are conjoined with what Howard calls the ergodic hypothesis: 'Regardless of its initial state, an isolated system will eventually visit every one of its microstates compatible with relevant macroscopic constraints' (Howard 2007, 146). So the question asked is whether or not macroscopic thermodynamic phenomena are indeed *emergent* rather than *resultant* from mechanical calculation (incidentally, both terms were coined by Lewes in the very same passage) (see Lewes).[8] It seems the answer is yes, and this is important to the extent that it undermines reductionist prejudice that trades on the success of intertheoretic reduction, which fails. In the case of non-ideal systems in statistical thermodynamics, the equations of the state used to estimate the energy of interaction between molecules cannot be deduced from any fundamental theory. A crucial factor is that of equilibrium, which is the central notion of thermodynamics, yet it is a macro feature. Take temperature, this only makes sense for a system in equilibrium, but it does not exist at the micro. We are therefore losing our 'cool', quite literally (see Lombardi and Labarca 2005, 131).[9]

Howard surmizes that the confusion arises because the understanding of particles has been wrongheaded; we think in too atomistic a manner. 'What we, today, call particle physics is, the name notwithstanding, not really a theory of particles' (Howard 2007, 155). Interestingly for this chapter, Lévy-Leblond employs the metaphor of a rainbow: particles have 'the mode of existence of rainbows' (see Bitbol 2008). No wonder, as a particle, properly understood, is but a manifestation of a quantum field; it is excitation of fields that affords the cardinality of subsets of particles to which they are in a one-to-one correspondence, not to mention permutation invariance wherein one particle can be replaced without cost to the

8. There is little doubt he was under the influence of Mill.

9. Howard's favourite example of the failure of reduction, and more precisely the scuppering of supervenience, is that of quantum mechanical entanglement, touched upon already, which is irreducibly holistic (it should be noted that we can understand such holism even in the absence of entanglement) (see Seevinck). Schrödinger who introduced the notion of entanglement called it 'not *one* but rather *the* characteristic trait of quantum mechanics' (Schrödinger 1935b, 555). On entanglement, see Schrödinger 1935a, 1936.

system (this being somewhat analogous to molecular turnover for organisms, when all our molecules are changed over time but we remain 'Susan' (see Chakravartty 2017, 139). In short, particles are themselves configurations, or patterns. In terms of reduction, condensed matter physics certainly enjoys explanatory autonomy because it can't be reduced, but again, not because of a failure of supervenience (if there is such a chimera).

It should be noted that a plague is visiting two houses, that of the reductionist and the emergentist:

> Deconstructing the formal concepts of substance and of property in Quantum Mechanics is precisely as challenging for the reductionist as it is for the supporter of 'true' emergence of high level intrinsic properties. The physical process may have no substantial roof of emergent properties, but it has no substantial ground of elementary properties either, according to the most straightforward reading of Quantum Mechanics. (Bitbol 2007, 302)

Yet a danger lurks here, one that would allow the return of the idea of a 'base', one Bitbol is at pains to point out, 'It is wrong to assert that Quantum Mechanics displays "ontological" emergence. What emerges is only a new mode of possible cognitive relation between the microscopic environment and the available range of experimental devices' (Bitbol 2007, 301). Appeal to Quantum Mechanics, is therefore, a matter of *epoché*, which is to say, it is to aid us in suspending our natural attitude (*die natürliche Einstellung*) to employ Husserl's phrase, but in so being must not itself assume the vacant place (Legion would then would be the new resident) (see Bitbol 2002, 202).

Another interesting example of an autonomous macro feature, one insensitive to micro constituents, is that of superconductivity. The point here is that superconductivity has all the micro constituents any reductionist would desire, as mentioned already, yet its defining features (infinite conductivity, flux quantization, the Meissner effect, for example), *transcend* any such base (see Teller 1992, 106). Pines and Laughlin point to this transcendence in terms of higher organizing principles, symmetry breaking especially. An example of such a principle is that of continuous symmetry breaking, that after all allows us 'to render exact the Josephson quantum, and localization, which is responsible for the quantum Hall effect' (Morrison 2012, 149).[10] Now, such transcendence means there is a strong sense of immunity to the 'small', that being the case they are what is referred to as 'model independent', that is, not causally linked to a microphysical base, as such things are a moveable feast, as it were (once again replacing fermions with bosons) (Morrison 2015, 106). When superconductivity is involved bosons crowd together, indeed they pile together to the point that they act as if they

10. Localization is the absence of diffusion in terms of waves that caused by a high concentration of defects or disorder in crystal or solids, which in terms of electric properties, and an appropriate solid, can turn conductors into insulators.

were one, and therefore as if there was no electrical resistance at all. It should be noted that due to what is known as Cooper pairing two electrons (fermions) act, or better become a single boson (this pairing is the result of being submerged in a sea of opposite-charged particles). Consequently, fermions act as a boson, and bosons then can act themselves as a single entity. Characteristics to be found in a boson (even though composed of fermions) have nothing in common with their constituents, indeed they stand in contradiction. Again, what is crucial in understanding a system is not its elements, change as they do, continually, but the bonds of that system, which are ineliminable. As Morrison says, 'symmetry breaking (here the breakdown of elector magnetic gauge invariance) provides the explanation of emergent phenomena but the specific microphysical details of how the symmetry is broken are not part of the account' (Morrison 2015, 106; see Morrison 2006, 881). 'You cannot make a magnet by putting dipoles together "one at a time". Cobbling is simply not available. You have to act on them all at once. That's what resheathing is: acting at a higher scale so as to contain a new state upon things at lower levels' (Thalos 2013, 97). This is analogous to Haaken's work on lasers, in terms of his notion of the 'enslaving principle'; indeed, it is analogous to biological systems *in general*.

Thalos gives the example of a painting, which is approached atomistically will permit only a version of a Pointillist painting, all that is macro is really an aggregation of the micro, but this is wrongheaded. First, the painting requires a suite of coordinated, mutual effort, from the materials to the painter. Second, any notion of the pointillist perspective being the real (here the micro) is wrong, for even if we were to allow this incoherent notion, any point, as it were, would become a painting of its own, and so on, downward, but crucially, up also. Kim argues that 'to cause any property (except those at the very bottom level) to be instantiated, you must cause the basal conditions from which it arises (either as emergent or as resultant)' (Kim 1999, 24). But surely, as Thalos asks, is this not simply sheer prejudice against the macro? Yes, is the very easy answer. Why can we not simply speak about bringing about the macro directly? (Thalos 2013, 99) After all, as Butterfield asks, 'Whoever said that ontology concerns only "supervenience basis", i.e., the putative set or level of facts that determine (subvene) all other facts? That is: there is plenty of scope for ontological discussion of supervening ("higher level") facts and theories.' Well, we can, indeed we do, no matter the overlay of micro language, for the micro is so underdetermined that it only splutters into talk when summoned by the macro: here it is the puppet, and never the master, for it's recruited, gang pressed even, and made to do or be something by way of participation. Yet we should point out that in some sense we can say that the macro is part of the micro, the converse also (see McGivern and Rueger 2010). Indeed, McGivern talks of a *mixture* of the micro and macro scales (McGivern and Rueger 2010, 389, fn. 12). 'We take the mixed solution seriously as a description of the behaviour of the system that is indeed composed of ingredients from both scales' (McGivern and Rueger 2010, 393). In other words, it is what it is only by what it relates to, and therefore is told what it is, it does not tell us; sure, what it can afford is a given potential, but its content and extent are not susceptible to any form of analytical closure. We can intervene analytically, which is to intervene on the 'parts'; with synthesis, by contrast, we intervene at the global level, and there we deal with the whole. These two must work in concert. For the former involves upward causation, the latter downward, but they are not in competition (see Bitbol 2012, 244). This point can be reinforced when we consider what it is that reductionism seeks to hide from us. Namely, the null possibility, in other words, there could be much less order, to the point of there being none. The reason this is hidden from us is because its possibility reveals, as we have suggested, that even the most basic element supposed by physicalism is a result of order, it is not just there – language such as base, with the inflection of minute size, so small as to not warrant explanation is mere assertion. But no, the very real possibility of nihilism scuppers the innocence or unproblematic ordinariness of the micro.[11]

Cooking with Chemistry

By moving our attention to chemistry, we see a possible escape from the shackles of both reduction and emergence. As we know already, a crucial phenomenon is that of symmetry breaking, which is essential as it induces or allows for the advent of structure and hierarchical levels, importantly, though, all such levels, and indeed subsystems cannot be subsumed by first principles; they cannot because they lack an ordering principle, which is to say, the hierarchy of theories has the structure of a non-Boolean lattice. As for symmetry breaking, think of a vertical pole on a flat plane, before any fall the pole has 360-degree symmetry, but if it does fall then this symmetry is broken and a new macroscopic feature emerges. Anderson applies this to many types of chemical molecules, for example glucose, which is a six-carbon sugar molecule that is synthesized in our cells. Now, such molecules are characterized by a handedness, what is called chirality (*cheir* means hand in Greek), and this gives rise to enantiomers or enantiomorphs. The simple point of Anderson's analysis is to bring attention to the fact that glucose involves a break in left-right symmetry, it does because glucose molecules are only ever right-handed, but in so being no physical laws are violated, yet these same laws cannot tell us why organisms enzymatically construct only right-handed sugars (or similarly why the amino acids that make up proteins are left-handed). Symmetry and structure are inversely correlated, for instance a snowflake has much greater structure, yet the water in a snowflake once dispersed throughout a room has much greater symmetry (Ward 2010, 261). This is a phenomenon that rides above and between different theoretical frameworks, it does because it functions as a structural constraint, and this is applicable to both low- and high-energy physical systems.

Morrison gives the example of the electroweak theory which postulates symmetry breaking by way of the Higgs mechanism that is said to explain bosonic

11. On the possibility of the type of nihilism meant here, see Coggins (2010) or Lowe (2002).

masses; in addition, superconductivity involves symmetry breaking, again via Cooper pairing. These facts of life cannot be deduced from physical laws; rather, they are emergent properties, being so in the strongest possible sense, but not in a bid to outrun reduction, there is no need, for as we said above they are not in opposition. Morrison characterizes the situation thus:

> The relation between ontological and epistemic independence is especially important since the latter is a necessary but not sufficient condition of emergence; the fact that we need not appeal to micro phenomena to explain macro processes is a common feature of physical explanation across many systems and levels. Instead, what is truly significant about emergent phenomena is that we supposedly cannot appeal to microstructures in explain or predicting these phenomena even though they are constituted by them. (Morrison 2015, 92)

Morrison's contributions to this area of debate are excellent, but I am tempted to disagree with the idea that the pervasiveness of epistemic independence in terms of the micro in relation to the macro means that the appellation emergence is withheld. For Morrison it must be that we cannot appeal to microstructure *tout court* that bestows the title. Crowther argues that emergence is better thought of in terms of novelty and autonomy – an autonomy that frees its manifestation from inclusion in a comparison, as such. Such novelty and autonomy are discernible by way of underdetermination and renormalization, and arises because of, for example, symmetry breaking.

For Crowther, avoiding conflation allows us to decouple reduction and emergence and this frees emergence up, thereby enabling it to be picked out on its own terms, which is to say it is not necessarily articulated in terms of a rejection or scuppering of reduction (see Butterfield 2011). That being the case its pervasiveness can be embraced. Such an embrace is made possible because we are not concentrating on the relation between levels, but rather on the ET itself, one that presents novelty and autonomy right before our eyes that need not stray elsewhere. As Crowther says, 'The positive conception of emergence as being simply novelty and autonomy of the low-energy physics compared to the high-energy physics means we can consider individual cases of emergence' (Crowther 2015, 439). As a result, we can consider emergence even before universality (it being only potentiality). Imagine having only one example of an emergent phenomenon, such as a superconductor, composed of a particular metal, well, Crowther argues, we would not be able to speak of it as emergent. Instead if we simply look at the phenomenon and what it entails the idea of reduction does not appear, nor even fail to appear, so speak. According to Crowther the gauge symmetry is all that matters here, and not the micro, but we are it seems forced to label such a phenomenon as being merely resultant. This is most interesting. Two quick points. We must ask is Crowther's point not retrospective, which is to say, she must be carrying the idea of reduction within the analysis, however unconsciously, otherwise why make the point: why pick out superconductivity or notice it. Crowther can escape this dilemma and reach her analytical observation if we abandon emergence and reduction as stable terms. After all, the simplicity of reduction invokes a complication all of its own, perplexity is even more forthcoming, or at least just as much. Of course, we might be worried that this leads to vacuity. But no, not if we have recourse to effectiveness as our approach, indeed tool of analysis. Both reduction and emergence tell us about the sheer richness of our common cosmos.

Carbon offers an intriguing insight into just how complicated yet ordered the reality we try to explain is. Carbon is tetravalent, which simply means that it can bind to four chemical groups. For example, in chloroform it is bound to one hydrogen atom and three chlorine atoms ($CHCl_3$), while in methane, carbon binds to four hydrogen atoms (CH_4). When it is bound to four different groups, as with natural α-amino acids, two different forms are made manifest, which are chiral: they are mirror images of each other but are not superimposable (see above). Yet only the L-form (left-handed) of natural α-amino acids is present in nature (exceptions are almost negligible). A compound bearing the same molecular formula, yet a different structure is called an isomer. And those with different spatial order are termed stereoisomers, and the two chiral mirror images are enantiomers (or optical isomers). When natural α-amino acids are synthesized a 50/50 mixture of the enantiomers is produced. But as said nature only employs the L-form, and that which shares the chirality it is referred to as homochirality (e.g. all proteins, natural sugars, and so on). If nature abhors a vacuum, she loves asymmetry. The ubiquity of L-form natural α-amino acids reduces the combination from trillions to a single protein having only one optical isomer, this allows the production of structure and order, otherwise nature would be something of a Buridan's ass, just with a near infinity of plates from which to choose, and nothing would follow from that. The asymmetry at the molecular level allows the manifestation of symmetry at the macro, thus there are things to be known in reality, or indeed there is an intelligible reality at all: A cosmos. But the story does not stop there, because although the macroscopic is symmetrical it too involves symmetry breaking, and the resultant asymmetry accommodates innovation. Asymmetry can be found throughout nature, for example the seemingly ubiquity of the helix, from plants to DNA, a wonderful explanation of this is the Fibonacci sequence (and its related golden sequence), which reveals how the self-similarity involved in phyllotaxis makes sense (a beautiful, conspicuous example of this is the shell of the Nautilus, a snail mollusc).

Domains

Echoing the point Cao made above, once again instead of levels we see that it makes sense to think in terms of enclaves of order, or domains, thus escaping any proprietary manner of speaking (Thalos 2013, 22; Humphreys 2016; Agazzi 2014). We do so because domains are more dynamic, overlapping, indeed mutually reinforcing, or as Humphreys would have it, neutral (metaphysically speaking, that is, not in bed with physicalism): Agazzi calls them regional ontologies (Humphreys

2016, 39, 121). Moreover, talk of domains allows us to speak of elements from one domain being reduced to another one, or elements of one domain emerging from another domain, but when one does so there is no implication of hierarchical ordering (Humphreys 2016, 122). If we cow-tow to physics as a master science we must face the simple dilemma: 'Physics cannot tell us why Biology exists' (Thalos 2013, 26). As we read above, Thalos advocates a scale-free model, which insists that difference in scale does not mean a difference in level, by now asking what a level would be, as we can't seem to find one. It is not that we don't rest on the shoulders of giants, we do, but at the same time those who we rest upon, we hold up, for there is no 'floor' (or base) beneath the feet of those giants. Indeed, the use of the word 'beneath' should be wisely set aside. According to Thalos, reductionism fails because supervenience is not fit for purpose, and emergentists half agree (as we know). Put differently, the ordinary is extraordinary at the same time as the extraordinary is ordinary. Once again, B traits supervene on A traits, such is the case that there cannot be a change in B traits without a change in A traits and from this comes the point (made most strongly by Kim), that surely, and applying Ockham's razor, why bother with the B traits at all, in terms of deeming them real, rather than epiphenomenal, and therefore lacking independent causal efficacy: mere shadows cast by the real stone. How do we even pick out B traits, to then reduce them, and how does any such process of dismissal get going on reductionist terms, for surely it is laced with higher/different order traction. Supervenience would mean that when A-characteristics are held fixed, B-characters will be fixed as well. But here's the rub: why can't we look at it from the other way around? That is, when B-characteristics are fixed so too are A-characteristics, and any notion of priority can only be smuggled in extraneously, that is, by way of presumption, not argument. This is suggestive, for once again we can begin to realize that reduction is not in opposition to emergence, nor the converse, yet both have borrowed logics, otherwise they are but placeholders. Downward and upward causation are not to be thought of as either real or illusory, rather, for Bitbol, they should be thought of as two modes of intervention that are indexed to their appropriate level of intervention.[12] According to Bitbol we should replace foundational, substantialist dualism with a functional duality of domains. (To some degree he shares this understanding with Humphreys and arguably Thalos.) Again, the problem with both reductionism and emergentism is that both presume a substantial dualism between base parts and relational wholes.

> The miracle is the only thing that happens, but to you it will not be apparent,
> Until all events have been studied and nothing happens that you cannot explain
>
> (W. H. Auden, *For the Time Being, Recitative*)

12. That being the case, they are not substantialist theses as they contribute to the very definition of their terms, therefore they are relational concepts (Bitbol 2012, 234).

It's All in the Mixt: Fusion Cooking

Chemistry operates with a completely different ontology than physics, in so doing the purported gap between the sciences and the humanities is bridged; not that there is a gap to be closed. Rather, different modes of engagement need to be recognized and each given their appropriate ontological weight. That being the case, chemistry serves as a major force to collapse the logic of reductionism and as a vehicle for the return of *scientia* and the abandonment of 'science' as a stand-alone term. (We should recall that the word 'scientist' was only coined as a term of art in 1833 by William Whewell, and was resisted until around the First World War.) Chemistry proves to be a hinge between the humanities, so-called, and the natural sciences. Kant famously said that chemistry was an *uneigentliche Wissenschaft* (though the context is often ignored), but after his so-called *Übergang* (or transition) he went so far as to say that 'philosophy always belongs to chemistry' (van Brakel 2013, 69–94). Chemistry is founded on a form of art – Man as *Homo faber*, indeed *Homo creator and Homo prospectus:* 'Everything in sight has the chemistry's art written into it' (Baird, Scerri and McIntyre 2006, 4). This forces us to rethink our scientific ontology *as a whole:* what nature entails will become more expansive, challenging the division between *techne* and *physis* – after all, Aristotle said that *physis* depended on art.

What, then, is this ontology? For a start, the dilemma of realism versus positivism is irrelevant; chemistry in a sense accepts neither. It is operationally realist, certainly, but the world of chemistry is more open, more conscious of the constitutive interventions of questions, experiments, and so on (see Bensaude-Vincent 2008, 52). But this does not mean it is less of a science, rather it is in this regard, more of a science. In addition, chemistry extends our understanding of reality, insofar as its ontology extends to incorporate the fabricated. A crucial point is that relation and plurality are the 'basis' of chemistry, so too processes, not some pretend kernel. This is important for several reasons. For the sake of space let us browse only a couple of these.

As we know, the microphysical, the apparent base, the one over which philosophy fawns does not even know *what it is or can do* without the macrophysical – it is blind and dumb, so to speak. Some examples: carbon, hydrogen and oxygen are not 'sweet', but $C_{12}H_{22}O_{11}$ (i.e. sucrose) is. As something of an addendum, only when such chemicals are taken up and metabolized does sucrose come into existence. Similarly, acidity is not reducible to a micro base, for it is a behaviour, one that is multiply realized. This idea that something is defined not by what it is, as it were, but what it does, or indeed is done to it goes back to Alan Turing and Alfred North Whitehead. One might think that alcohols are 'reducible' (they being a molecular group that bear a hydroxyl bond), that is true, but what different alcohols afford are radically different despite the same molecular base (isomers): take dimethyl ether and ethyl alcohol, both C_2H_6O, described by the same Hamiltonians (the measurement of the total energy in a system, in whichever form), yet their structures are represented as $(CH_3)_2O$ and C_2H_5OH – one gets you drunk while the other does not; once again, take C_3H_8O, this gives rise to three isomers, namely, methoxyethane, propanol and rubbing alcohol. So from where does the difference come? (Bensaude-Vincent 2008, 214; see Hendry 2017, 149) It seems it must be generated because of the relevant molecular structures (see below). Unsurprisingly, quantum mechanics has a problem with isomers (a problem if one is employing it for reductive purposes) as the same wave function is applied despite the different molecular structures. When any base is used, it is *used up* in the formation of the whole. Here we are back with Plato and Aristotle's 'mixt'. As Nordham points out, 'the problem of mixt has never been resolved or displaced' (Nordham 2013, 731). Take Hamiltonian operators, which are not just resultants but also configurational, the former being but a special case: reduction requires all Hamiltonians to be merely resultant. As Aristotle says, 'To the eye of the Lynceus nothing would be combined' (Aristotle, *De generatione*, I.10, 328a13f.). Consequently, the 'mixt' as he called it would be a fiction, or only apparent. If there was to be a true *mixt* then a new stuff had to emerge, one in which constituents were no longer there in actuality, or (to put it in cruder terms), but only in potentiality (see Aristotle, *De generatione*, I.10, 327b23ff.). Millenia later, Pierre Duhem captures the either/or condition of a *mixt* well: 'What in general, then, is a mixt? Some bodies, the one different from the other, are brought into contact. Gradually they disappear, they cease to exist, and in their place a new body is formed, distinguished by its properties from each other of the elements, which produced it by their disappearance. In this mixt, the elements no longer have any actual existence. They exist only in potentiality, because on destruction the mixt can regenerate them' (Duhem 2002, 5–6). The biologist Young brings to your attention a telling situation. He argues that the essence of a living thing is that it consists of atoms caught up into a living system and made part of that whole. In so being it disappears – a *mixt*, in our terms (see Young 1971, 86–7). Think of when a sperm joins an ovum, they both cease to be, and a new entity is produced. Or take Aquinas' notion, following Aristotle, that there are three types of soul: vegetative, animal and rational. These three combine, and therefore mix, in so doing a person is created, there now being only one soul, which is the substantial form of the body. This is analogous to Humphrey's idea of fusion, which is 'meant to show how supervenient relations fail to apply when fusion occurs, that role is generalizable to any similar dependency relations' (Humphreys 2016, 72). What does he mean by fusion? Quite simply, and again echoing Plato and Aristotle's notion of *mixt*, (micro) parts when combined, here fused, no longer exist, so we cannot speak about a base at all – the micro disappears.

Consolidating the above insights, we should understand that a molecule plus its environment can be thought of as a supersystem, and such a system appears to have the power to break the symmetry of its subsystems thereby generating structure, as all symmetry breaking appears to (see above), yet any such power does not appear to arise from any of the supersystem's subsystems. Surely, Hendry asks, is this not downward causation? The Coulomb Schrödinger equations describe mere assemblages of electrons and nuclei rather than molecules, which are structured entities. (This is illustrated, as we know, by the fact that isomers, which are distinct molecules sharing the same molecular formula, share the same Coulomb Schrödinger equation.) (Hendry 2010b, 186). 'Molecular structures cannot be recovered from the Coulomb Schrödinger equations, but not because of any mathematical intractability. The problem is that they are not there in the first place' (Hendry 2010a, 213). According to Vemulapalli the applications of quantum theory to chemical bonding involve calculations of quantities that have neither representation nor status in the original theory. Hence, 'many extra-quantum assumptions must be made to adapt the basic theory to the form that is useful in chemistry' (Vemulapalli 2006, 196). These assumptions surely confirm Wooley's worry that molecular structure as a 'universal attribute in molecular science is ... not securely founded in quantum theory'. Consequently, it must be put in by hand (Wooley 1978, 1007; Wooley 1988, 56). 'The same event can have a chemical or physical description ... but no privileged description exists' (van Brakel 2000, 171).

This lack of privilege, the God's-eye view, becomes evident, at least it does in relation to chemistry, when we realize that if quantum mechanics turned out to be wrong, molecular shape, chirality, and so on would persist, quite happily. The relation, therefore, between physics and chemistry is not asymmetrical, but rather symmetrical (see van Brakel 2000, 177). To use acid as an example: Acidity as a function seems to make sense in terms of affordance.[13] An acid is a substance that produces carbon dioxide when it comes into contact with calcium carbonate or sodium; it also turns phenolphthalein, which is red, colourless. So, an acid is realized by being acetic acid, or sulphuric acid, the property of being hydrochloric acid, and so on, but where, we might ask, is the unifying principle, for all such realizers are disparate. However, the Lewes concept of acid seems to challenge the above.[14] According to this account, an acid is an electron-pair acceptor. Consequently, sulphuric acid, hydrochloric acid, and so on qualify as an acid because they bear this property of accepting an electron pair. This common feature provides unification for a range of functionally characterized substances by way of this physical property. That being the case we cannot really speak of emergence; indeed, it is more accurate to speak of reduction: chemical property (being an acid) = physical property (electron-pair acceptance). Moreover, can we not now question the functional characterization of being an acid? I'm not sure. I wonder if the same functionalization does not simply reappear at a different level, or more accurately in a different way. That which qualifies as being acidic taking the physical property as decisive seems odd, insofar as they are very different, yet they are realizing the same behaviour. Also, sulphuric acid is characterized in terms of electron-pair acceptance, does this mean that all such acceptance leads to the same result?

> The symmetry problem impacts on ontological reduction via its commitment to the completeness of physics in two ways. The first is direct – if the acidic behaviour of the hydrogen chloride molecule is conferred by its asymmetry, and the asymmetry is not conferred by the molecule's physical basis according to physical laws, then surely there is a prima facie argument that ontological reduction fails. (Hendry 2010b, 187)

13. A neologism coined by J. J. Gibson in 1979.

14. There are also accounts of acid such as that offered by Arrhenius and Bronsted-Lowry conceptions.

Prevalent in much thinking on mereology is that of Armstrong's idea of an *ontological free lunch*, in short, 'mereological wholes are identical with all their parts, taken together Symmetrical supervenience yields identity' (Armstrong 1997, 12). (He apparently does not understand the fusion involved in every *mixt*.) This is not true for chemistry; indeed, it is not true for much. For example, it does not apply to most chemical combinations. What is implicit in Armstrong's thinking is the idea that once you have identified the 'individuals' then when they are thought to be in a system they will remain the same 'individuals', but this is not even close to reality. Take the example of hydrocarbons. Earley makes the simple point that even when analysing relatively small hydrocarbons, by way of NMR spectroscopy, what is revealed and is revealing is that there are many types of hydrogen atoms to be found and in addition various types of carbon atoms all of which exhibit different properties (see Earley 2006, 213). In short, combination is transformation. But we are bewitched, once again, by a false picture: 'being identical molecule for molecule = identity'. Van Brakel asks,

> Does any such identity include the velocities and relative position of the molecule? If it does, then it undermines the idea of macroscopic objects being identical 'molecule for molecule' because the velocities and relative positions of the molecules are constantly changing. If it does not, then, say, temperature would be excluded as a relevant macroscopic parameter for two objects being (in)discernible. At the macroscopic level everything is statistical and changing – hence no two things are ever the same'. (van Brakel 2000, 79)

Van Brakel admits that such differences are averaged out, approximately at the macro level, sure, but that is not the level so desired by the microphysical fundamentalist.

To break with this bewitchment, as mentioned, it is crucial that we alter our conceptual image of essences, especially of the notion of microphysical essence, for any underlying essence, so-called, varies with context, indeed it is variable as nominal essences. Returning to the question of water once again it is important, as von Brakel makes clear, that we realize there are $H_3O\cdot$ and OH^- ions in liquid water. There are H_4O_2- molecules as well as other H_2O polymers in water vapour. Moreover, temperature and other contextual variables dictate how much ionization or dimerization and polymerization there is. Water can be thought of then, pragmatically as H_2O, in terms of necessity, but never in terms of sufficiency (Van Brakel 2000, 80). In short, water is not simply H_2O (see Chang 2014). Once again, that is because there is no one microphysical structure or essence, or at least essence as it is normally understood; context and function have an enormous part to play in an understanding of such materials, or 'stuff'. Luisi provides another telling example, that of the aromaticity of a benzene molecule. The point being here that any such property is not, of course, there in the atoms of that molecule. But more importantly, 'when a benzene molecule is created, the orbitals of the carbon atoms and those of the hydrogen are changed; the molecule as a whole affects the properties' (Luisi 2002, 183–200). The molecule constrains its parts, certainly in this case, in terms of motion. Therefore, there is emergence, and our lunches start to look anything but free.

As the story we are often told, goes, 'A great deal of classical chemistry has been reduced to atomic physics', that is, the contention of Oppenheim and Kemeny (1956, 7). Hans Primas is having none of it. He responds by pointing out how quickly reduction is scuppered by quantum mechanics: 'According to our present understanding of the first principles of quantum mechanics, atomism as well as reductionist-mechanistic philosophies have no longer any scientific basis … . Modern quantum mechanics compellingly demands a *multitude of context-dependant* descriptions. Atomism is dead, the material world is certainly not constituted out of independently existing elementary systems' (Primas 1991, 165).

Tellingly, mechanics is in fact very much against supervenience. As Thalos points out, 'According to analytical mechanics, macroscopic structures – whether imposed externally or simply known to hold – are treated as limitations on the total number of dynamically micro-possibilities, from "above" as it were.' Indeed, mechanics allows microquantities to serve as manifestations of degrees of freedom. Such mechanics refrains from introducing quantities, such as forces, that mediate between the actual and the possible, hence it leaves gaps, doing so in the absence of strictures as to how they should be filled (Thalos 2013, 85). Crucially, mechanics takes the independence of boundary conditions as fundamental, and therefore as priority, hence not derivative. A major consequence of such analysis is summed well by Lanczos, 'The analytical approach to the problem of motion is quite different [from the Newtonian approach]. The particle is no longer an isolated unit but part of a "system". A mechanical system signifies an assembly of particles which interact with each other. The single particle has no significance; it is the system as a whole which counts' (Lanczos 1949, 4).

The art, the alchemy that chemistry both is and reveals (biology also), can be better noticed when we realize that reality is so dynamic, as mentioned already.[15] Returning to the question of microstructures helps. Matter is always restless it is never

15. Interestingly, New Mechanists such as Bechtel and Craver do not embrace reductionism, insofar as any notion of deductive-nomological reduction is rejected, and there is a constant appeal to multi-level explanations arguing that mechanisms occur in nested hierarchies (see Machamer, Darden and Craver 2000). Such pluralism is evident in chemistry, as we shall see, but in biology also. Biology now employs many different approaches, such as evolutionary developmental biology, systems biology, synthetic biology, epigenetics (whereby traits are inherited from the environment and passed on not via the genotype, crudely put) and so on. Or again, biology not only deals with the survival of the fittest (a biology of becoming), but the arrival of the fittest also (a biology of being). There is, so to speak, both a theatre (structure, or being) and play (phylogeny and survival). The reason for this being is that if there is to be biology at all, and not merely the flux of phylogeny, which is itself nonsensical, then such metaphysically rich approaches are required. Under the cosh of reductionism biology loses its domain, including the organism. We now realize that quite the opposite is happening.

quiescent. Indeed, as quantum mechanics has demonstrated the dynamic nature of matter, just as relativity theory shows that matter cannot be separated from its activity. (In this sense Francisco Suarez was incorrect for not even God could create pure matter, that is, a matter wholly devoid of form, empty matter. Aquinas insisted God could not, unlike Suarez, something for which Aquinas was condemned for asserting in 1277, in Paris.) There is stability in matter, but it is more like what Ludwig von Bertalanffy coined *Fließgleichgewicht* ('flux equilibrium'). The neglect of matter's dynamism is of course apparent in microphysical fundamentalism.

Size matters: It should be brought to the attention of our imagination that 'if we decrease the size of particles of the same chemical substance down to the nanometer scale, their stuff properties begin to vary at a certain size. Moreover, in that size range the properties also vary with the shape of the particles. It appears at the nanometer scale *form* philosophy takes over' (Schummer 2008, 16; emphasis mine). Put simply, the smaller the particle 'the more atoms are on its surface compared to the number of bulk atoms; and surface atoms behave differently from bulk atoms and differ in their behaviours depending on surface curvature' (Schummer 2008, 16).[16] The point being that molecular structure is not enough to classify in a safe and useful way. For example, toxicity is a disposition, of course, but one that only occurs by way of interaction. Take Gold: at bulk level it is non-toxic, yet gold nanoparticles are cytotoxic – it all depends on *size*. Thalidomide provides another example of such variance in terms of affect. Thalidomide usually consists in two optical isomers, and these are mirror image structures, which offer a very subtle difference. The problem being is that one of these is an effective treatment for morning sickness while the other causes all the dreadful birth defects we have come to associate with its name. So, we are back with Plato's *pharmakon*, wherein we cannot decide whether something is poisonous or beneficial. Once again, the microphysical is 'dumb and blind', they know not what they do. As Paracelsus put it, generalizing this ambiguity to all substances: *Alle Dinge sind Gift, und nichts ist ohne Gift; allein die Dosis macht, daß ein Ding kein Gift ist* ('All things are poison, and nothing is without poison; only the dose permits something not to be poisonous'). We are in the *mixt*; we are a *mixt*: a composite of form and matter, as Aquinas would say, or in tripartite terms: body, soul and spirit being so as one person.[17] Consequently, in terms of *mixt*, there can be no body either *simpliciter*. Indeed, 'the mystery of body is no less deep than that of mind. Therefore, there

16. This is the sort of conundrum faced by *REACH* (Registration, Evaluation and Authorization of Chemicals), that which is used by the European Community.

17. Interestingly, Philo of Alexandria, Flavius Josephus and St Paul all writing at the same time argued for just such a tripartite anthropology, a mixt. Of course, this was taken up by most of the church fathers such as St Irenaeus: 'For that flesh which has been moulded is not a perfect man in itself, but the body of a man, and a part man. Neither is the soul itself, considered apart by itself, the man; but it is the soul of a man, and a part of a man. Neither is the spirit of a man, for it is called the spirit, and not a man; but the commingling and union of all these constitutes the perfect man' (*Against Heresies*, VI, I).

is no provision in physical science for how to count bodies, even once we have produced an inventory of quantities or the features making an appearance within a certain spacetime' (Thalos 2013, 81). Aristotle would have agreed, doing so because he realized that we are a mixt: 'There is no part of an animal that is purely material or purely immaterial' (Aristotle, Part. An. 1.3643A, 24–6). Thus Plato warned, 'We ought not to seek the good in the unmixed life but in the mixed one' (Plato, *Philebus*, 61b).

Conclusion: A Bun in the Oven

For Plato, Aristotle and Aquinas following them, it is true to say that *omnis scientia bona est* ('all knowledge is good'). This idea must, however, be set over and against the vice of *curiositas*. Consequently, knowledge must be accompanied by the virtue of *studiositas*.[18] Clement of Rome wrote a letter to the Corinthians, it being the oldest datable document of Christianity, and in it we find the following: 'The head is nothing without the feet, and so the feet are nothing without the head … but all conspire (*panta sympnei*) and are united in their subordination to the task of preserving the whole body.'[19] All breathe together (Greek *sympnei*, Latin *conspirant*). The originally medical language is extended beyond biology to anthropology, but still further, for the *sympnoia* of the parts is meant, then, to communicate the universe entire (the Cosmos, which derives from the Greek 'κόσμος' for order) as a *sympnoia pantōn*, one underwritten by a shared *skopos*, or *telos*, which is the *eros* of all knowledge, *scientia*. Crucially, according to Aquinas, God is 'the final end of the whole universe' (SCG, I,I). As a philosopher of science puts it, if all knowledge is good, it is because it involves a form of faith summoned by desire. 'If the realization of the real involves participation in the real, the separation of mind and world as separate spheres is no longer possible and with that, there is also no separation between the superficiality of mere appearances and the depth of explanatory structures behind the appearances' (Needham 2012, 737). It should be remembered that when in the Hebrew Bible we are told that 'Adam knew his wife' (*Genesis* 4.1), the term 'knowing' is the same for sexual intercourse and the cognitive act. This Hebrew word for knowing is *jadah*. As Aristotle says, 'For the mind somehow is potentially what it thinks (*ta noēta*)' (*De anima*, III 4, 429B30-31). Again, 'Knowledge thus activated is thus the same as the thing' (*De anima*, III 7, 431B21). Thus 'the mind that is active is the objects' (*De anima*, III 7, 431B17). And the soul, the *psuchē*, the mind entails a *capax omnia* – for the soul is 'somehow all things.'

With regard to this desirous faith, Augustine implores us, 'believe so that you may understand' *(crede, ut intelligas)*. Aristotle concurs: 'Some trust is necessary

18. The term *curiositas* was coined by Cicero. The Latin term stems from the Greek words *periergos* and *polupragmon* (see Torchia 2013).

19. Clement, I Epist. Ad Corinth, 37.4. This echoes a passage from the Hippocratic book – *Peri trophés* 23: 'One confluence, one conspiration, all in sympathy with one another!'.

for whoever wants to learn *(dei pisteuein ton manqanonta)*' (*Sophistical Refutations*, 2, 165b3). *Scientia* realizes that all sciences are subalternate, that is, they live by way of borrowed logics of which they cannot give an account. As Aquinas pointed out, theology as *sacra doctrina* must borrow from revelation (e.g. knowledge of the Trinity), likewise physics borrows from mathematics, or the example given by Aquinas is that of optics. Once again, this is the simultaneous moments of the cataphatic and the apophatic which all knowledge entails. Therefore, all *scientia* is a *marriage* of discourse, a concert of effort that forfeits imperial ambitions, for *scientia* seeks engagement, not subsumption, nor eradication. It is a marriage that does not seek the purely isolated, or the reduced. Maybe the person of *scientia* is imbued with what the poet Keats called 'negative capability' (Shakespeare was one person he had in mind), 'when man is capable of being in uncertainties, Mysteries, doubts, without any irritable reaching after fact & reason' (Keats 1958, 193). This is reminiscent of Rilke's reading of Orpheus: 'Song, as you teach it, is not covetousness, or the quest for something one might fully obtain. Song is existence' (Rilke 2015, 29). Our common universe, such an uncommon song.

References

Agazzi, Evandro, *Scientific Objectivity and Its Context* (Dordrecht: Springer, 2014).

Anderson, Philip W., *'More is Different'*, *Science*, New Series, 177, no. 4047 (4 August 1972): 393–6.

Armstrong, David, *A World of States of Affairs* (Cambridge: Cambridge University Press, 1997).

Baird, David, Eric Scerri and Lee McIntyre (eds), 'Introduction', in *Philosophy of Chemistry: Synthesis of a New Discipline* (Dordrecht: Springer, 2006), 3–18.

Balthasar, Hans Urs von, *Presence and Thought: An Essay on the Religious Thought of Gregory of Nyssa*, trans. M. Sebanc (San Francisco: Ignatius Press, 1995).

Batterman, Robert, *The Devil in the Details: Asymptotic Reasoning in Explanation, Reduction and Emergence* (Oxford: Oxford University Press, 2002).

Batterman, Robert, 'Autonomy of Theories: An Explanatory Problem', *NOûS* 00, no. 0 (2017): 1–16.

Batterman, Robert, and S. Green, 'Biology Meets Physics: Reductionism and Multi-scale Modelling of Morphogenesis', *Studies in History and Philosophy of Biological and Biomedical Sciences* 61 (2017): 20–34.

Bensaude-Vincent, B., 'Chemistry Beyond the "positivism vs Realism" Debate', in *Stuff: The Nature of Chemical Substances*, ed. K. Ruthenberg and J. van Brakel (Würzburg: Königshausen & Neuwmann, 2008).

Berry, Michael, 'The Bakerian Lecture: Quantum Chaology', in *Dynamical Chaos*, ed. M. V. Berry, I. C. Percival and N. O. Weiss, vol. 186 (Royal Society of London: Princeton University Press, 1987), 183–98.

Berry, Michael, 'Asymptotics, Singularities and the Reduction of Theories', *Logic, Methodology and Philosophy of Science* IX (1994), 597–607.

Bertolaso, Marta, *Philosophy of Cancer* (Dordrecht: Springer, 2016).

Bitbol, Michel, 'Science as if Situation Mattered', *Phenomenology and the Cognitive Sciences* 1 (2002): 181–224.

Bitbol, Michel, 'Ontology, Matter and Emergence', *Phenomenology and the Cognitive Sciences* 6 (2007): 293–307.
Bitbol, Michel, 'Is Consciousness Primary?' *NeuroQuantology* 6, no. 1 (2008): 53–72.
Bitbol, Michel, 'Downward Causation without Foundations', *Synthese* 185 (2012): 233–55.
Bodin, Jean, *Colloquium Heptaplomeres de rerum sublimium arcanis abditis* (Paris/London, 1857; 1970: New York: Olms, 1970).
Borges, Jorge-Luis, *The Aleph*, trans. A. Hurley (London: Penguin Books, 2000).
Brakel, J.-V., *Philosophy of Chemistry: Between the Manifest and Scientific Image* (Leuven: Leuven University Press, 2000).
Bulgakov, Sergei, *Philosophy of Economy*, trans. C. Evtuhov (New Haven and London: Yale University Press, 2000).
Butterfield, Jeremy, 'Emergence, Reduction and Supervenience: A Varied Landscape', *Foundations of Physics* 41 (2011): 920–59.
Cao, Tian Yu, *Conceptual Developments of 20th Century Field Theories* (Cambridge: Cambridge University Press, 1997).
Cao, Tian Yu, 'Ontological Relativity and Fundamentality: Is QFT the Fundamental Theory?' *Synthese* 136 (2003), 25–30.
Cao, Tian Yu, and S. Schweber, 'The Conceptual Foundations and the Philosophical Aspects of Renormalization Theory', *Synthese* 97 (1993): 33–108.
Casellani, Elena, 'Reductionism, Emergence, and Effective Field Theories', *Studies in History and Philosophy of Modern Physics* 33, no. 20 (2000): 251–67.
Chakravartty, Anjan, *Scientific Ontology* (Oxford: Oxford University Press, 2017).
Chang, H., *Is Water H2O?* (Dordrecht: Springer, 2014).
Chibbaro, S., L. Rondoni and A. Vulpani, *Reductionism, Emergence, and Levels of Reality: The Importance of Being* Borderline (Dordrecht: Springer, 2014).
Ciauncia-Garrouty, A., 'Relations Between Levels From a Chemical Perspective', in *The Philosophy of Chemistry: Practices, Methodologies and Concepts*, ed. J.-P. Lloyd (Newcastle: Cambridge Scholars Press, 2013), 558–74.
Coggins, Geraldine, *Could There Have Been Nothing?* (New York: Palgrave, 2010).
Condillac, Étiene, *La Logique ou l'art de rasionner (1780)*, new edition (Paris: Vrin, 1981).
Córdoba, Mariana, and Olimpia Lombardi, 'A Kantian Response for the Philosophy of Chemistry', in *The Philosophy of Chemistry: Practices, Methodologies, and Concepts*, ed. J.-P. Llored (Newcastle upon Tyne: Cambridge Scholars Publishing), 478–90.
Crowther, K., 'Decoupling Emergence and Reduction in Physics', *European Journal of Philosophy of Science* 5 (2015): 424–42.
Cunningham, Conor, *Genealogy of Nihilism: Philosophies of Nothing and the Difference of* Theology (London and New York: Routledge, 2002).
Duhem, Pierre, *Mixture and Chemical Combination and Related Essays*, trans. and ed., P. Needham (Dordrecht: Kluwer, 2002).
Earley, Joseph Emmet, 'Some Philosophical Implications of Chemical Symmetry', in *Philosophy of Chemistry*, ed. D. Baird, E. Scerri and L. McIntyre (2006), 207–20.
Garnett, Campbell, 'Scientific Method and the Concept of Emergence', *The Journal of Philosophy* 39 (1942): 477–86.
Georgi, H., 'Effective Quantum Field Theories', in *The New Physics*, ed. P. Davies (Cambridge: Cambridge University Press, 1989).
Georgi, H., 'Topics in Effective Theories', in *Advanced School on Effective Theories*, ed. F. Cornet and M. Herrero (Singapore: World Scientific, 1995), 88–122.
Hartmann, Stephan, 'Effective Field Theories: Reductionism and Scientific Explanation', *Studies in History and Philosophy of Modern Physics* 32 (2001): 267–304.

Hawking, Stephen, *Is the End in Sight or Theoretical Physics?* (Cambridge: Cambridge University Press, 1980).
Hendry, Robin F., 'Is There Downward Causation in Chemistry?', in *Philosophy of Chemistry: Synthesis of a New Discipline*, ed. D. Baird, E. Scerri and L. McIntyre (Dordrecht: Springer, 2006).
Hendry, Robin F., 'Chemistry: Emergence vs. Reduction', in *Emergence in Mind*, ed. C. Macdonald and G. Macdonald (Oxford: Oxford University Press, 2010), 205–21.
Hendry, Robin F., 'Emergence vs Reduction in Chemistry', in *Emergence in Mind*, ed. C. MacDonald and G. MacDonald (Oxford: Oxford University Press, 2010).
Hendry, Robin F., 'Reduction, Emergence and Physicalism', in *Philosophy of Chemistry: Handbook of the Philosophy of Science*, ed. A. Woody, R. F. Hendry and P. Needham (Amsterdam: Elsevier, 2012).
Hendry, Robin F., 'Prospects for Strong Emergence', in *Philosophical and Scientific Perspectives on Downward Causation*, ed. A. Paoletti and F. Orila (New York and London: Routledge, 2017).
Howard, Don, 'Reduction and Emergence in the Physical Sciences', in *Evolution and Emergence*, ed. N. Murphy and W. Stoegers S.J. (Oxford: Oxford University Press, 2007).
Humphreys, Paul, 'How Properties Emerge', *Philosophy of Science* 64 (1997): 1–17.
Humphreys, Paul, *Emergence: A Philosophical Account* (Oxford: Oxford University Press, 2016).
Jackiw, Roman, 'The Unreasonable Effectiveness of Quantum Field Theory', in *Conceptual Foundations of Quantum Field Theory*, ed. T. Y. Cao (Cambridge: Cambridge University Press, 1999), 148–59.
Keats, John, *The Complete Letters of John Keats, 1814-1821*, ed. H. E. Rollins, vol. I (Cambridge, MA: Harvard University Press, 1958).
Kemeny John, and Paul Oppenheim, 'On Reduction', *Philosophical Studies* 7 (1956): 6–19.
Kemp, Martin, *Seen/Unseen: Art Science and Intuition from Leonardo to the Hubble Telescope* (Oxford: Oxford University Press, 2006).
Kim, J., 'Making Sense of Emergence', *Philosophical Studies* 95: 10 (1999): 3–36.
Kronz, Frederick, and Justin, Tiehen, 'Emergence and Quantum Mechanics', *Philosophy of Science* 51 (2002): 324–47.
Lanczos, Cornelius, *The Variational Principle of Mechanics* (Toronto: University of Toronto Press, 1949).
Laughlin, Robert, and David Pines, 'The Theory of Everything', *Proceedings of the National Academy of Sciences* 97 (2000): 28–31.
Laughlin, Robert, and David Pines, 'The Theory of Everything', in *Emergence: Contemporary Readings in Philosophy and Science*, ed. M. Bedau and P. Humphreys (Cambridge, MA: MIT Press, 2007), 259–68.
Lewes, George Henry, *Problems of Life and Mind*, vol. ll (London: Trübner & Co., 1874/1875).
Lombardi, Olimpia and Martin Labarca, 'The Ontological Autonomy of the Chemical World', *Foundations of Chemistry* 7 (2005): 125–48.
Lowe, Jonathan, 'Metaphysical Nihilism and the Subtraction Argument', *Analysis* 62 (2002): 62–73.
Luisi, Pier Luigi, 'Emergence in Chemistry: Chemistry as the Embodiment of Emergence', *Foundations of Chemistry* 4 (2002): 183–200.
Machamer, Peter, Lindley Darden and Carl Craver, 'Thinking about Mechanism', *Philosophy of Science* 67 (2000): 1–25.
McGivern, Patrick, 'Reductive Levels and Multi-Scale Structure', *Synthese* 165 (2008): 53–75.

McGivern, Patrick, and Alexander Rueger, 'Hierarchies and Levels of Reality', *Synthese* 176 (2010): 379–97.
Meissner, U. G., *Effective Field Theories of the Standard Model* (Singapore: World Scientific, 1992).
Morrison, Margaret, 'Emergence, Reduction, and Theoretical Principles: Rethinking Fundamentalism', *Philosophy of Science* 73 (2006): 876–87.
Morrison, Margaret, 'Emergent Physics and Micro-Ontology', *Philosophy of Science* 79 (January 2012): 141–66.
Morrison, Margaret, 'Complex systems and Renormalization Group Explanations', *Philosophy of Science* 81 (2014): 1144–56.
Morrison, Margaret, 'Why is More Different?, Introduction', in *Why More is Different: Philosophical Issues in Condensed Matter Physics and Complex Systems* (Dordrecht: Springer, 2015).
Nagel, Ernst, 'The Meaning of Reduction in the Natural Sciences', in *Science and Civilization*, ed. R. C. Stauffer (Madison, WI: University of Wisconsin Press, 1949), 99–135.
Nagel, Ernst, *The Structure of Science* (London: Routledge and Keegan Paul, 1967).
Needham, Paul, 'Aristotle's Theory of Chemical Reaction and Chemical Substances', *Philosophy of Chemistry*, ed. D. Baird, E. Scerri and L. McIntyre (Dordrecht: Springer, 2006), 43–67.
Needham, Paul, 'Kant's Legacy for the Philosophy of Chemistry', in *The Philosophy of Chemistry*, ed. J.-P. Llored (Newcastle: Cambridge Scholars Press, 2013), 69–94.
Needham, Paul, 'Metachemistry', in *The Philosophy of Chemistry*, ed. J.-P. Llored (Newcastle: Cambridge Scholars Press, 2013), 725–43.
Nelson, David, *Defects and Geometry in Condensed Matter Physics* (Cambridge: Cambridge University Press, 2002).
Neurath, Otto, 'Sozologie im Physikalismus', *Erkenntnis* 2 (1931): 393–431.
Nickles, T., 'Two Concepts of Intertheoretic Reduction', *Journal of Philosophy* 70, no. 7 (1973): 181–201.
Nordham, Alfred, 'Metachemistry', in *The Philosophy of Chemistry*, ed. J.-P. Llored (Newcastle: Cambridge Scholars Press, 2013), 725–43.
O'McGarity, Thomas, and Wendy Wenger, *Bending Science* (Cambridge, MA: Harvard University Press, 2010).
Oppenheim, Paul, and Hilary Putnam, 'The Unity of Science as a Working Hypothesis', *Minnesota Studies in Philosophy of Science* 2 (1958): 3–36.
Primas, Hans, 'Reductionism: Palaver without Precedent', in *The Problem of Reduction in Science*, ed. E. Agazzi (Dordrecht: Kluwer, 1991).
Proclus, *Commentary on Plato's Parmenides*, trans. G. Morrow and J. Dillon (Princeton, NJ: Princeton University Press, 1979).
Ricoeur, Paul, *From Text to Action*, trans. K. Blamey and J. B. Thompson (Evanston, IL: Northwestern University Press, 1991).
Rilke, Rainer Marie, *Sonnets to Orpheus*, trans. D. Polikoff (New York: Angelico Press, 2015).
Sarkar, Sahotra, *Genetics and Reductionism* (Cambridge: Cambridge University Press, 1998).
Schaffer, Jonathan, 'Is There a Fundamental Level?', *Noûs* 37 (2003): 498–517.
Schaffer, Jonathan, 'Monism: The Priority of the Whole', *Philosophical Review* 119, no. 1 (2010): 31–76.
Schrödinger, Erwin, 'Die gegenwärtige Situation in der Quantenmechanik', *Die Naturwissenschaften* 23 (1935a): 807–12, 823–8, 844–9.

Schrödinger, Erwin, 'Discussion of Probability Relations Between Separated Systems', *Proceedings of the Cambridge Philosophical Society* 31 (1935b): 555–662.

Schummer, Joachim, 'Matter versus Form, and Beyond', in *Stuff: The Nature of Chemical Substances*, ed. K. Ruthenberg and J. van Brakel (Würzburg: Königshausen & Neuwmann, 2008).

Sklar, Lawrence, 'Types of Inter-theoretic Reduction', *British Journal for the Philosophy of Science* 18 (1967): 109–24.

Sklar, Lawrence, 'I'd Love to be a Naturalist – if Only I Knew What Naturalism Was', *Philosophy of Science* 77 (2010): 1121–37.

Teller, Paul, 'A Contemporary Look at Emergence', in *Emergence or Reduction? Essays on the Prospects of Nonreductive Physicalism*, ed. A. Beckerman, H. Flohr and J. Kim (Berlin: de Gruyter, 1992), 139–53.

Thalos, Mariam, *Without Hierarchy: The Scale Freedom of the Universe* (Oxford: Oxford University Press, 2013).

Torchia, Joseph, *Restless Mind: Curiositas and the Scope of Inquiry in St Augustine's Psychology* (Milwaukee: Marquette University Press, 2013).

Truesdell, Clifford, *An Idiot's Fugitive Essays on Science: Methods, Criticism, Training, Circumstances* (Dordrecht: Springer-Verlag, 1984).

van Brakel, Jaap, 'The Nature of Chemical Substances', in *Of Minds and Molecules: New Philosophical Perspectives on Philosophy of Chemistry*, ed. N. Bushan and S. Rosenfield (Oxford: Oxford University Press, 2000).

van Brakel, Jaap, 'Philosophy of Science and Philosophy of Chemistry', *Hyle* 20 (2014): 11–57.

van Fraassen, Bas, *The Scientific Image* (Oxford: Oxford University Press, 1980).

Vemulapalli, Krishnah, 'Physics in the Crucible of Chemistry: Ontological Boundaries and Epistemological Blueprints', in *Philosophy of Chemistry*, ed. D. Baird, E. Scerri and L. McIntyre (Dordrecht: Springer, 2006).

Ward, Peter, 'What Is the Diversity of Life in the Cosmos?', in *Water and Life: The Unique Properties of H2O*, ed. R. Lynden Bell, S. Conway-Morris, J. Barrom, J. Finney and C. Harper (London and New York: CRC Press, 2010), 249–58.

Weinberg, Steven, 'Newtonianism, Reductionism and the Art of Congressional Testimony', *Nature* 330 (1987): 433–7.

Weiss, Nigel O., *Dynamical Chaos*, vol. 186 (Royal Society of London: Princeton University Press, 1987), 183–98.

Weisskopf, Victor F., 'In Defence of High Energy Physics', in *Nature of Matter: Purposes of High Energy Physics*, ed. L. C. L. Youan (Brookhaven National Laboratory, 1965), 33–46.

Wilson, Kenneth, 'Renormalization Group and Critical Phenomena. I. Renormalization Group and the Kadanoff Scaling Picture', *Physical Review*, B 4 (1971): 3184–205.

Wimsatt, William, 'Reduction and Reductionism', in *Current Research in the Philosophy of Science*, ed. P. Asquith and H. Kyburg (East Lansing: Philosophy of Science Association, 1979), 352–77.

Wippel, John, *The Metaphysical Thought of Thomas Aquinas* (Washington: CUA Press, 2000).

Wooley, Guy, 'Must a Molecule Have a Shape?', *Journal of the American Chemical Society* 100 (1978): 1073–8.

Wooley, Guy, 'Must a Molecule Have a Shape?' *New Scientist* 120 (1988).

Young, John, *An Introduction to the Study of Man* (Oxford: Clarendon, 1971), 53–7.

Part II

INTERACTIONS

Chapter 5

SUSTAINABILITY: INTERACTION BETWEEN SCIENCE, ETHICS AND THEOLOGY

Robert S. White
Cambridge University

Introduction

The importance of sustainable living when there are only finite resources available is a no-brainer. Planet Earth is self-evidently a finite resource, and most people would agree that we should not overuse its provision. Yet once the resource is much larger than our own local environment, we find it hard to change behaviour to limit our use of that resource. An example is global climate change. Many people agree with the scientific evidence that global climate change is occurring, yet make little change to their lifestyles to reduce their own carbon footprint. When asked about this, they are likely to say that their own actions have only a trivial impact in the light of global problems, so there is no point. Ethical imperatives are much more likely than scientific evidence to encourage individuals to make changes to their lifestyles for the sake of others. Theological insights from the Judeo-Christian religion point to a creator God who revels in his creation, and who wishes his people to care for it in a way which caters for the good of others and which gives glory to Him (Moo and Moo 2018, 144). Ultimately, ethical and theological drivers are more likely to engender change in the lifestyles of individuals and communities than are raw scientific facts.

Sustainability: The Long View

If by sustainability we were to mean that there is essentially no change in the environment, then it is clear that planet Earth has never had a long-term unchanging environment. Rather, it has evolved continually from its formation some 4,560 million years ago to the present day. But if we mean by sustainability the ability to sustain a rich diversity of life, then Earth has been a uniquely successful place for doing so over its long history. Remarkably, the surface temperature has remained between 0°C and 100°C for almost all its existence, which means that the surface water (of which there is a finite supply on the planet) has never completely frozen, nor boiled off into space. That in turn has allowed life to thrive. And thrive it has, with evidence for the presence of simple cells on Earth from almost as soon as it was possible for life to exist. Through a long evolutionary history, life developed into multicellular forms, then in a great explosion around 600 million years ago to the diverse range of animals of which we are one of the descendants.

Indeed, it may be that the interaction and feedbacks between life and the planet is what has maintained a habitable, though changing environment. This is often called the Gaia hypothesis, named after the goddess who personified the Earth in Greek mythology. This idea was suggested in 1972 by James Lovelock (Lovelock 1972, 579–80) and subsequently developed in collaboration with Lynn Margulis (Lovelock and Margulis 1974, 2–10). The teleology suggested by such an interpretation has sat uncomfortably with scientific explanations of evolutionary processes, and more recent studies have suggested that there is unlikely to be a direct link between the evolution of organisms and a drive to habitability, as suggested by the Gaia hypothesis (Tyrrell 2013, 209). Nevertheless it is undeniable that organisms change the environment by their very act of living, a fact which has been taken to *extremis* by modern humans as I discuss below.

Geological history also shows that at the level of species there is little long-term stability even in the absence of humans. Over 99 per cent of the species that have ever lived on Earth are now extinct. That is estimated as around 5,000 million species extinct. In the normal course of events there is a steady turnover of species. This has been punctuated by several extreme extinction events. Perhaps the most famous is the extinction at the Cretaceous–Tertiary boundary some sixty-six million years ago. This marked the end of the dinosaurs and opened the way for mammals, of which humans are one species. The extinction was probably triggered by a large asteroid impact at Chicxulub on the Yucatán Peninsula in Mexico, although at the same time there was massive and widespread volcanism which created the Deccan Traps in India covering 500,000 square kilometres: so the Earth suffered a double whammy, both of which probably contributed to the mass extinctions.

Sustainability: The Human Timescale

Humans are now by far the largest modifiers of Planet Earth. Every year we move thirty-five billion tons of rock and soil, which is more than all the ongoing geological processes put together. In the latter half of the twentieth century nearly one-third of the arable lands worldwide were so heavily affected by water erosion (the biggest proximate cause of soil loss), wind erosion or chemical and physical degradation that they had to be taken out of cultivation. Compared with the Earth's normal 'background' rate of extinction, the current rate is estimated to be anywhere from 100 times to 1,000 times higher as the result of human activity.[1] In

1. See www.whole-systems.org/extinctions.html

total, there are today one-third fewer wild animals on the planet than there were just forty years ago (Lynas 2011, 32). Only 3 per cent of terrestrial vertebrate flesh is wild. Of the rest, one-third is human and two-thirds is domesticated animals. Every single year we are extracting and releasing into the atmosphere carbon which took a million years to accumulate: we have raised greenhouse gas levels in the atmosphere to levels not experienced on Earth for over three million years, causing rapid global climate change at a rate never before experienced by humans. There are many more examples of the impact of humans on the Earth. Indeed the changes are so great and so abrupt on a geological time scale that a new geological epoch, the Anthropocene, meaning 'Age of Humanity' has been proposed.[2]

One of the enduring features of humans is their ability to invent technological aids which make life easier, or in other words allow more energy to be spent on activities above mere survival (though it should be acknowledged that for a significant number of people alive today they do still have to spend most, if not all of their waking hours simply getting enough food to eat or water to drink). The technological advances started with basic activities such as making weapons and hand tools to assist in catching wild prey and dismembering them. The next big step came in the development of cooking, which enabled humans to get far more energy out of food than eating it raw. Then about 10,000 years ago came animal husbandry and agriculture, which allowed humans to settle in villages, towns and cities as they took charge of the production of food in the surrounding region. This allowed people to have a more reliable source of food, as well as storing food for use during times of the year when it could not be harvested. It is striking that Earth's climate has remained remarkably stable over the last 10,000 years since the end of the last Ice Age for all but the last few decades: this is what enabled settled agriculture and arguably the development of civilization to develop and prosper.

Selective breeding, both of plants and animals, allowed a steady increase in the human population. Then, most recently, the ability to produce inorganic fertilizers and to capture the energy from burning fossil fuels led to the so-called "Green Revolution" with its huge increase in crop production. Largely as a result of the availability of synthetic fertilizers, combined with the use of fossil fuels to power machinery and the development of high-yielding hybrid crops, the 'Green Revolution' of the mid-twentieth century led to a huge increase in the amount of food the world could produce. Between 1950 and 1984, global grain production increased by a factor of more than 2.6 (Kendall and Pimentel 1994, 198-205). It was this 'Green Revolution' that staved off a chronic shortage of food that would otherwise have resulted from the fourfold increase in population over the same period. But it carried many negative environmental effects, including soil erosion, pollution of ground and surface water and public health problems caused by pesticides. Some commentators are now looking to the next technological fix to bypass some of those deleterious side effects. Maybe it will be genetic modification to allow crops to be more drought resistant, to grow in more extreme climates or

2. This term was first proposed by Crutzen and Stoermer (2000, 17–18).

to produce higher yields. However, ongoing use of genetically identical crops over wide areas (monocultures) carries with it the increased risk of catastrophic failure from such things as unforeseen disease.

Technological Fixes to Unsustainability?

It is tempting to believe that new technological fixes are always just around the corner, and will save us from whatever looming disaster lies ahead. They have to some extent done so in the past, as with the 'Green Revolution' discussed above. But there are almost always unintended consequences which make reliance on an as yet untested or even unknown technology extremely dangerous. For example the term 'Green Revolution' is rather a misnomer. It depended primarily on the massive use of fertilizers, pesticides and fossil fuels. All of these carry deleterious effects that are now beginning to be felt. Synthetic fertilizers and the burning of fossil fuels now add 160 million tons of nitrogen to the environment every year and humans are destined to double the turnover rates not only of the terrestrial nitrogen cycle but also of the nitrogen cycle of the entire Earth. That has led to eutrophication (enrichment of nutritious elements) of terrestrial and aquatic systems, global acidification and stratospheric ozone loss (Gruber and Galloway 2008, 293-6).

Underpinning almost all advances in the way humans use new technology to interact with the environment is energy (Pain 2017, 135–7). For most of human history people relied on their own muscle power, fuelled by food. This was supplemented by the use of animals once they had been domesticated. The first main inanimate power source was the use of water mills, starting around the third century BC and becoming widespread across Western Europe by the end of the eleventh century. But those who controlled access to water rights controlled the power, so it was not an egalitarian development. The spread of windmills, first seen in seventh-century Persia and reaching Europe around AD 1150 gave more people access to power, though it was subject to the vagaries of the weather. The next major change began in England in the sixteenth century, when coal was used both for industry and for heating. Along with the invention of the steam engine, coal powered the Industrial Revolution, initially in Britain and then across the world. When electricity generation was developed in Europe and the United States in the 1880s, centralized coal burning with distributed electricity fostered the rapid spread of both industrial and domestic machinery and appliances. Nuclear power, solar photovoltaic and wind generated electricity entered the mix in the late twentieth century. But the first big game changer was oil.

The first commercial oil well was drilled in Pennsylvania, United States, in 1858–9. By 1886 the first motor cars driven by internal combustion engines were built. The high energy density of oil rapidly made it an energy source of choice for cars, ships and aeroplanes. The amount of energy produced by an engine burning one gallon of petrol produces as much work as about one month's labour of a person. Mechanization powered by fossil fuels is an astonishingly powerful tool. In terms of the energy we each utilize in our daily lives, either personally or by proxy in the foods we eat and the resources we consume, people in high-income countries like the United States and the United Kingdom are relying on the work of the equivalent of 200 human labourers. The consequence of our dependence on machinery in modern agriculture is that so much fossil fuel is used in food production, packaging and distribution that it is estimated that 7 to 10 kilocalories of energy are expended for every kilocalorie of food consumed in the United States. We are, as one recent writer has said, essentially eating fossil fuels (Pfeiffer 2006). This can't continue indefinitely: we are probably close to the peak in conventional oil supplies, although there is sufficient coal to last at least another century. Hydrocarbons are such complex polymers that they can be used for a huge range of plastics and synthetic materials that are important in all walks of daily life. Indeed in the future it might well be seen as a crime that humanity simply burned such a valuable polymer: it's rather like burning the family heirloom Chippendale furniture just to keep warm for a few hours one evening.

Apart from the diminution of such valuable resources, the burning of fossil fuels has another arguably even more criminal side effect: it causes global climate change by the release of greenhouse gases into the atmosphere. This has been posited as the biggest danger facing humanity in coming decades. Already heatwaves, floods, extended droughts and extreme weather are claiming tens of thousands of lives every year. Even with the seemingly modest present-day increase of about 1°C in global average temperature since the mid-twentieth century, some 30 per cent of the world's population is currently exposed for at least twenty days per year to a threshold in air temperature and relative humidity above which increased deaths occur. By 2100 this percentage is predicted to increase to nearly 50 per cent, even if drastic reductions in greenhouse emissions are achieved globally. If emissions continue to increase as they currently are doing, then three quarters of the world's population will be exposed to this deadly threshold. It is almost inevitable that global temperature rises will cause an increased threat to human life, but this will be greatly aggravated if greenhouse gases are not reduced substantially (Mora et al. 2017, 501–6).

Among the so-called natural disasters, including earthquakes and volcanic eruptions, the most mundane of them, namely flooding, is actually the deadliest of all (White 2014, 207). And an increased risk of flooding is often a direct result of global climate change caused by humans releasing greenhouse gases by burning fossil fuels.

Let's return to the question of whether new technology will get us out of the problem of global climate change. The idea that we can simply carry on with 'business as usual', however damaging our actions, because one day in the future we will find a magic bullet to fix global climate change is simply an excuse not to change our comfortable ways. What we need to do is the much harder and at least initially more costly option of finding ways to continue the improvements in living standards that have undoubtedly accompanied technological progress but without irrevocably damaging the planet (although note that the eminent economist Lord Sterne has argued that a small investment today will save a lot more money in the years to come – and that is on a purely financial basis without even considering the cost in human lives and well-being) (Stern 2007).[3]

A partial solution is for humanity to wean itself off its addiction to oil and to develop alternative energy sources from renewable sources like wind, sun and tides. There is some progress in that direction in high-income countries. There is less in low-income countries, which argue that the lifestyles we enjoy in Europe and North America were bought by the burning of fossil fuels in the past, so they deserve to enjoy the same improvement in living standards as we have inherited. This brings us directly to the ethical issues surrounding sustainable living.

Ethics

It is demonstrably the case that the effects of global climate change cause more suffering to the poor and the disempowered in low-income countries than to those in high-income countries. As an example, as already mentioned floods are the most deadly 'natural disaster'. But they disproportionately affect low-income countries and poor people, because high-income countries can protect themselves more effectively. Following tidal surges in 1953 which inundated large areas of eastern England and the Netherlands, killing 2,190 people, both the British and the Dutch built major sea defences. The Thames barrier and the Eastern Scheldt surge barrier which were built as a result are the largest moveable flood barriers on Earth, costing billions of dollars to build (in 2018 prices). But the thirty million largely subsistence farmers living within 1 metre of sea level in the deltaic region of Bangladesh simply don't have the financial resources to attempt to build such barriers, even if it were technically feasible to do so in such an area. Present rising sea levels are caused mainly by global climate change resulting from the burning of fossil fuels in high-income countries. But people in high-income countries are not usually the ones who suffer the worst consequences – the people most affected played little part in causing the problem in the first place, and have least resources to cope with it.

The link between poverty and vulnerability to disasters is a moral and ethical issue for those of us who live in high-income countries. This is because, first, we have technological and scientific understanding of the causes and probabilities of disasters that if shared properly with vulnerable societies could hugely reduce the casualties that occur when events such as floods happen. People in high-income countries also have access to real-time data such as that from satellites which means that they can predict weather patterns and watch floodwaters developing long before the flood surges reach towns and villages downstream. But often expensive solutions are not required. Teaching about ways of irrigating land that

3. Subsequently in 2008 Lord Stern commented that his report had underestimated the risks of climate change and the damage associated with it, which made the likely costs of inaction even greater than reported in his original report (see Adam 2008).

does not waste water or of using drought-resistant crops is a way to help vulnerable rural populations. There are many examples where education is a key factor and this is one area where people from high-income countries can readily assist.

A second more potent reason is that of global climate change. Those of us in the high-income countries who have benefitted from burning cheap fossil fuels, thereby causing global climate change have a moral duty to help those in low-income countries, who largely are the people who suffer from climate change. They need help to adapt to the inevitable changes that result. By and large, those in low-income countries neither caused the problem nor benefitted from the increase in standard of living bought with widespread fossil fuel usage. Many disasters are related directly or indirectly to climate change, including heat waves, floods, droughts, landslides and changes in weather patterns that impact agriculture and may lead to famines (Spencer and White 2007, 245). So a two-pronged response is appropriate. Even if greenhouse gases were (totally unrealistically) reduced to zero overnight, the Earth would continue warming for centuries to come because such a lot of heat is stored in the oceans and it will take centuries for that to work its way to equilibrium. It takes many decades to remove excess greenhouse gases from the atmosphere and sequester them in plants, soil or oceans. So meanwhile an ethical response would be for rich nations to help those suffering the effects of climate change by enabling them to mitigate and adapt to those changes, while also working to reduce their ongoing greenhouse gas emissions.

In addition to the question of present-day equity to people living elsewhere on the planet, from changes caused by past and present greenhouse gas emissions, there is the question of intergenerational equity for our children and grandchildren. If we consume non-renewable resources, accelerate the death of species on Earth and cause major changes to the climate system, then we are being grossly unfair to them. The then prime minister of Britain Margaret Thatcher enshrined this view in a speech to The Royal Society in 1988. She said, 'We do not hold a freehold on our world, but only a full repairing lease. We have a moral duty to look after our world and to hand it on in good order to future generations.'[4] This moral duty was subsequently formulated as a matter of principle in the Government White Paper *This Common Inheritance* (1990, 10). In the light of the pressure for continuing development to improve quality of life and well-being, the Brundtland pronouncement of 1987 was that 'sustainable development is development that meets the need of the present without compromising the ability of future generations to meet their own needs' (World Commission on Environment and Development 1987, 44). Can such sustainable development be sustained? Often it is hijacked by political pressures by governments for 'growth' in their economic development, which can leave the impression that the present trumps the future (Blowers, Boersema and Martin 2012, 1–8). Environmental changes usually occur gradually over long periods, whereas democratic governments are generally elected for relatively short terms of four to five years. So it can be difficult for

4. Quotation from Margaret Thatcher in a speech to The Royal Society in 1988.

governments to put in place policies that cost money and have an immediate impact on people (such as raising taxes on petrol), when the pay-off will not be for many years or decades to come. Perhaps the main constraint on this is that democratic governments rely on the populace to re-elect them. So if there is sufficient groundswell of opinion that some particular issue or policy matters to the population at large, even if there is a short-term cost, politicians are likely to respond to it because above all they want to stay in power.

Lastly, it is important to note that the interaction between the many facets of the environment and human activities are complex and non-linear, so we cannot always fully understand the consequences of our actions, even if the broad features are clear. Non-linear interactions may cause us to cross tipping points from which there is no return. An example would be complete melting of the summer Arctic sea ice, which has already lost more than half its area since satellite observations began in 1979. Since ice reflects sunlight, whereas water is much darker and absorbs heat, this produces a positive feedback. As the ice melts, the water heats up more than it otherwise would have done, and this can cause a runaway effect. The loss of the sea ice acts as an amplifier which makes the temperature increase even higher. It is estimated that if the planet were to warm up by an average of 2°C, the Arctic region would warm up by about 5°C. A tipping point such as this happens when a small perturbation is capable of driving rapid change. It is possible that we have already crossed the critical threshold for losing the Arctic sea ice.

A range of other possible tipping points have been discussed by a group of senior climate researchers,[5] but given the complexity of the interactions between the atmosphere, the climate and the surface of the Earth, there may well be others that have not yet been recognized. The possibility of tipping points with major uncontrollable consequences means that it is sensible to take a precautionary approach in our plans, particularly where climate change is concerned. As in many decisions in life, there may well be actions that we could take, and are able to take given our socio-economic and technical circumstances, but that we eschew for the sake of others. This brings us to another perspective which interacts with these decisions and our world view of how to make such decisions, which is the theological aspect.

Theology

In this section I concentrate on Christianity, although other religions often have generally similar views on the practices that we should seek to follow from religious imperatives (see Bell, Chaplin and White 2013). The foundation of a theological perspective on sustainability is based on the three-way relationship between God the creator, his creation and us his creatures. From the very beginning of the Bible, it is clear that the whole of the created order was brought into being by God and

5. A review of possible tipping points is provided by Lenton et al. (2008, 1786–93).

that he was pleased with what he had made. Although humans are special in that they are made 'in the image of God', we are still intimately connected to the rest of the biological order through our evolutionary history. The material world clearly matters dearly to God, so there is no sense that we should spiritualize our lives to the extent that we should believe that the material world is to be shunned, or is less important in God's sight. This is a first step in connecting theological views to finding ways of sustainable living.

The very first command given by God to humankind was to 'be fruitful and multiply and fill the earth and subdue it and have dominion over the fish of the sea and over the birds of the heavens and over every living thing that moves on the earth' (Gen. 1.28). Some have read this as a command to sustainable living: it is that, but it is rather more as well. It includes the necessity of having dominion over creation. This carries connotations of having to keep the Earth in good order, of having to work to do so and of using our understanding of how the natural order works to, for example, contain or eradicate disease. We are not just to sit back and assume that we can live in some notional idyllic pastoral setting without involving ourselves in intervening to care for it. Humans have been set the task as God's image bearers of caring for his creation as vice-gerents in its governance on his behalf. Part of that task includes using our scientific understanding of the world not just to care for the material world, but also to care for other people and other living things. Our ability to understand the created order is part of God's grace to us: We do not live in an unpredictable, chaotic world, but in a world upheld moment by moment by God's word. As the Apostle Paul wrote concerning Jesus: 'For by him all things were created … all things were created through him and for him. And he is before all things, and in him all things hold together' (Col. 1.16-17).

The early founders of the Royal Society, which arguably was when modern science was first instituted, had a clear view that by understanding the natural world better they could help the lot of humankind. Indeed the 1663 charter of the Royal Society declared explicitly that the society's studies 'are to be applied to further promoting by the authority of experiments the sciences of natural things and of useful arts, to the glory of God the Creator, and the advantage of the human race'. For many of the early fellows of the Royal Society this dual aim of giving glory to God by using scientific understanding for the good of humankind fitted exactly with their religious and ethical perspective of the world (White 2012, 157–79).

How then should we care for creation? There are many images in the Bible of God caring for his creation with all its complex interactions which can act as a model for us. For example, he doesn't do so just for the sake of humankind. God 'water[s] a land where no one lives, an uninhabited desert, to satisfy a desolate wasteland and make it sprout with grass' (Job 38.26–27). According to Ps. 147.9, God provides food not just for cattle but 'for the young ravens when they call'. Psalm 104.21 tells us that when lions roar, they seek their food from God. God also cares about predators in his world.

For Christians, the inherent value of the non-human creation derives in the first instance from God's pronouncements in Genesis that all that he has made is 'good'. If God values what he has made (even before the appearance of humankind), then we ought to value it too. The 'intrinsic' value of a created thing – of something that owes not just its material make up but its very being and even possibility of existence to a creator who 'calls into being things that were not' (Rom. 4.17; Heb. 11.4) – is rooted in the value that is bestowed upon it by that creator. Its value cannot be taken away by another mere creature. A biblical theology of creation, as well as a biblically-derived environmental ethos, is rooted in the affirmations in Genesis and many other passages in both the Old and New Testaments that it is God alone who establishes, upholds and sustains all of creation.

There is a direct causal link drawn in Scripture between the breakdown of the relationship between humans and their creator God and the breakdown of our proper relationship with the Earth (Marlow 2008). The Old Testament prophets warned that human evil and injustice could lead to environmental degradation and prevent the land being as fruitful and bounteous as God intended it to be. Often this is expressed as the land 'mourning' (e.g. Isa. 33.9; Hos. 4.3; Jer. 12.4), an idea that Paul picks up in the New Testament to describe the entire creation as longing to be set free from its 'bondage to decay' and its 'groaning as in the pains of childbirth' because of human sinfulness (Rom. 8.21-22). The Bible warns us against the folly of presuming that the Earth is limitless or that God would not allow us to suffer the consequences of our poor treatment of his creation. But the Bible also sets out clearly the sure and certain hope that we have in Christ for restoration and a setting of all things right in the new creation.

The Christian gospel is that the broken relationships between people and God are restored by the death and resurrection of Jesus, and that in the fullness of time the return of Christ will usher in a new creation where there is no more sin, no more death, 'no more mourning nor crying nor pain' (Rev. 21.4). The theologian Christoph Schwöbel comments that 'Jesus' story discloses the character of God's relationship to his creation as one by which God maintains his relationship to creation through the discontinuity of death' (Schwöbel 2000, 117; See also O'Donovan 1986). Though we can't understand the physical basis of the new creation, there will evidently be some continuity in that the best of human creativity will continue on into it: 'The kings of the earth will bring their glory into it' (Rev. 21.24). In addition to shady streams there will be a city, the new Jerusalem. And cities are products of human ingenuity.

Yet in the meantime we live in that in-between period between Christ's first coming and his return, a period of 'inaugurated eschatology'. Biblical hope in the new creation is not a promise of 'pie in the sky bye and bye'. Instead, it challenges Christians to faithful, righteous living that embodies God's promises in the here and now. That includes the imperative to live sustainably, even as we see all around us the crisis of environmental degradation (White 2009; Spencer and White 2007, 245). The Biblical vision of a fruitful creation broken by human selfishness comports frighteningly well with what science is coming to realize about humanity's outsize role on the Earth today. The Christian hope gives a solid incentive to work in this world today to put right some of those wrongs, because there is a long-term future for creation, and because we live in those in-between times where we ought by our lives attempt to live out those values of the Kingdom of God modelled by Jesus in his life. The title of a book I co-authored summarizes this well: *Hope in an Age of Despair: The Gospel and the Future of Life on Earth* (Moo and White 2013, 224). But the revised title of the same book published in the United States is perhaps a better, more uplifting summary: *Let Creation Rejoice: Biblical Hope and the Ecological Crisis* (Moo and White 2013).

Creation care, and the imperative to live sustainably, is inseparable from the Christian gospel (Bell and White 2016, 350). Christian theology has a powerful part to play in dialogue with scientific and ethical concerns about sustainable living. In many areas the ethical concerns of those with other religious beliefs, or with none, align closely with those derived from a Christian world view. Christian theology provides a firm metaphysical underpinning to endeavours to promote sustainable living: it points to the inherent worth of the material world in its own right, rather than treating it in an instrumental way as a means for fulfilling our own desires; it legitimizes the proper use of science and technology to understand and to use the natural world for the good of humankind; it points to the need for self-sacrifice or restraint for the sake of others, such as strangers suffering from the effects of climate change on the other side of the world, or as yet unborn future generations; and its insistence that 'the earth is the Lord's and all that is in it' (1 Cor. 10.26)[6] provides a perspective that sustainable living is not just an optional extra but a core part of the proper response of people to the world in which they live.

References

Adam David, 'I Underestimated the Threat, Says Stern', *The Guardian* (18 April 2008), available at www.guardian.co.uk/environment/2008/apr/18/climate change.carbonemissions

Bell, Colin, Jonathan Chaplin and Robert White (eds), *Living Lightly, Living Faithfully: Religious Faiths and the Future of Sustainability* (Cambridge: The Faraday Institute for Science and Religion, 2013).

Bell, Colin, and Robert S. White (eds), *Creation Care and the Gospel: Reconsidering the Mission of the Church* (Peabody, MA: Hendrickson, 2016).

Blowers, Andy, Jan Boersema and Adrian Martin, 'Is Sustainable Development Sustainable?' *Journal of Integrative Environmental Sciences* 9 (2012): 1–8.

Crutzen, Paul J., and Eugene F. Stoermer, 'The "Anthropocene"', *IGBP Global Change Newsletter* 41 (2000): 17–18.

Gruber, Nicholas, and James N. Galloway, 'An Earth System Perspective of the Global Nitrogen Cycle', *Nature* 451 (2008): 293–6.

Kendall, Henry W., and David Pimentel, 'Constraints on the Expansion of the Global Food Supply', *Ambio* 23 (1994): 198–205.

Lenton, Tim et al., 'Tipping Elements in the Earth's Climate System', *Proceedings of the National Academy of Science* 105 (2008): 1786–93.

6. See also Deut. 10.14; 1 Chron. 29.11; Neh. 9.6; Ps. 24.1; Ps. 104.24.

Lovelock, James E., 'Gaia as Seen through the Atmosphere', *Atmospheric Environment* 6 (1972): 579–80.

Lovelock, James E., and Lynn Margulis, 'Atmospheric Homeostasis by and for the Biosphere: The Gaia Hypothesis', *Tellus, Series A. Stockholm: International Meteorological Institute* 26 (1974): 2–10.

Lynas, Mark, *The God Species: How the Planet Can Survive the Age of Humans* (Washington DC: National Geographic, 2011).

Marlow, Hilary, *The Earth Is the Lord's: A Biblical Response to Environmental Issues* (Cambridge: Grove Books, 2008).

Moo, Douglas J., and Jonathan A. Moo, *Creation Care: A Biblical Theology of the Natural World* (Grand Rapids, MI: Zondervan, 2018).

Moo Jonathan A., and Robert S. White, *Hope in an Age of Despair: The Gospel and the Future of Life on Earth* (Leicester: Inter-Varsity Press, 2013).

Mora, Camilo et al., 'Global Risk of Deadly Heat', *Nature Climate Change* 7 (2017): 501–6.

O'Donovan, Oliver, *Resurrection and the Moral Order: An Outline for Evangelical Ethics* (Grand Rapids, MI: Eerdmans, 1986).

Pain, Stephanie, 'Power through the Ages', *Nature* 551 (2017): S135–7.

Pfeiffer Dale A., *Eating Fossil Fuels: Oil, Food and the Coming Crisis in Agriculture, Gabriola Island* (Gabriola Island, BC: New Society Publishers, 2006).

Schwöbel, Christoph, 'The Church as a Cultural Space: Eschatology and Ecclesiology', in *The End of the World and the Ends of God*, ed. John Polkinghorne and Michael Welker (Harrisburg, PA: Trinity Press International, 2000), 107–23.

Spencer, Nick, and Robert White, *Christianity, Climate Change and Sustainable Living* (London: SPCK, 2007).

Stern, Nicholas, *The Economics of Climate Change: The Stern Review* (Cambridge: Cambridge University Press, 2007).

Tyrrell, Toby, *On Gaia: A Critical Investigation of the Relationship between Life and Earth* (Princeton: Princeton University Press, 2013).

UK Government, The Stationary Office, *White Paper: This Common Inheritance* (London: HMSO, 1990).

White, Robert S. (ed.), *Creation in Crisis: Christian Perspectives on Sustainability* (London: SPCK, 2009).

White, Robert S., 'Take Ten: Scientists and their Religious Beliefs', in *Wisdom, Science and the Scriptures*, ed. S. Finnamore and J. Weaver (Oxford: Regents Park College, 2012), 157–79.

White, Robert S., *Who Is to Blame? Nature, Disasters and Acts of God* (Oxford: Lion Hudson, 2014).

World Commission on Environment and Development, *Our Common Future* (Oxford: Oxford University Press, 1987).

Chapter 6

About Continuous Creation, and Some Ethical Principles for Ecology

Fabien Revol
Catholic University of Leon

Introduction

The theme of biodiversity is very much connected with the issue of the current ecological crisis. It is, according to the French philosopher Virginie Maris, an expression born in the 1980s within a context of reflection on the protection of diversity of living beings (Maris 2010, 11). It has become one of the most important ecological issues alongside climate change. In his encyclical letter *Laudato si*, Pope Francis comes back quite often to this topic, giving it a strong theological meaning. For him it is a place for the display for God's creativity in the world, and also a place where God expresses his glory through different forms of creatures (§33). This reflection on the diversity of creatures in creation is intended to convey the idea that creatures have intrinsic value in existing (§69).[1]

This biodiversity proceeds from the interplay of interactions of interdependence, that is from ecological relationships. Here this means that species brought together build an ecosystem just as elements build a whole. But the ecosystem in its entirety is the reason for the existence of such elements as species in its midst. The diversity of being flows from the interactions present in an ecosystem through time, according to the evolutionary paradigm of biology.

Starting from this central idea of ecological relationship, new ways of doing ethics have been envisioned. The likes of Aldo Leopold already emphasized the idea of natural community according to the concept of 'land ethics'. This is an ethic acknowledging the true importance and meaning of the interdependence and interconnection of all natural beings in an ecosystem (Leopold 2001, 171). Its leads us to understand every natural being as part of a community of life, with its own and proper value of existence. Even before the well-known works of Leopold, there were French socialists in the early twentieth century who developed the concept of solidarity, in particular from their understanding of ecological relationships

1. For a discussion of the moral value of creatures, see Persson, this volume.

of interdependence in nature (Bourgois 2008, 23).[2] For Léon Bourgeois, the solidarity between beings, and among them, between humans and non-humans, is an ontological fact preceding any ethical choice. The recognition of such networks of interdependence calls for human responsibility in respecting this ontological solidarity (73). This is a strongly ecocentric claim that views the human being as an ecocitizen (Beau 2015, 307–9) that can be extended to a cosmic citizenship, assuming that this ontologic solidarity goes back to the produced matter in the heart of stars.

Considering the biosphere as a community of solidarity however implies a strong change in the way in which we speak of natural beings on earth, the way in which we understand them and even see them as beings. To fully accommodate this idea has implications for ethical reflection in both theology and philosophy. The protection of nature involves many of our images of the living world, and thus becomes a question of representation. In the 1960s, historian of medieval technologies Lynn White Jr raised the issue of images and representations of nature as being at the origin of the present ecological crisis. According to White, if such representations are not correct, they afford bad behaviour towards the natural environment on the part of human beings. This especially concerns Christian ethics because White's critique was addressed to Christian representations of nature and the place of the human being within it (White 1967, 1203–7). It is however possible to consider White's critique in terms of its positive implications: religious traditions and speech formulate representations of creation that may influence the mentalities and behaviour of believers. It strongly indicates a close link between the theology of creation and ecological ethics, insofar as theology of creation is the appropriate form of religious discourse for providing cosmological elements in an organized manner. This then is the theology of creation that serves as a source for catechesis and preaching to churchgoers and for Christian education. It thus sheds light on the responsibility of theologians and church leaders regarding ecology. What should be their contribution to such mental representations? How should they formulate them? How should they communicate them to their congregations, for instance in their preaching or material available for preaching, or in a context of Christian initiation?[3]

This chapter aims at providing philosophical and theological reflections within a Christian framework in order to seek the true meaning of biodiversity for a humanity responsible for the stewardship of the earth. With this purpose, I propose a strong link between the theological concept of continuous creation

2. Bourgeois speaks of a 'general law of reciprocal dependance' (*loi générale de dépendance réciproque*), 63. See also the works of Pierre Leroux: 'Human life … is attached to an unceasing communication with his brethren and with the universe' (1985, 129).

3. I am not inventing something new here. I simply intend to place myself in the stream of the Harvard initiative on religion and ecology that took place in the late 1990s, producing a series of large volumes on different religious traditions and ecology. See for instance Hessel and Ruether (2000).

and the need for the protection of biodiversity (see Revol 2013; 2015; 2017). This may provide complementary bases for ecological ethics, as in my view biodiversity can be interpreted as the main natural sign of a divine continuous creation; the understanding of its working and its values derivative from it can thus offer new principles for ecological ethics.

To this end I will reflect on the meaning of diversity in philosophy of nature and in theology, then present how the concept of biodiversity is connected to that of continuous creation, and eventually propose four new principles for an ecological ethic: creativity, diversity, vulnerability and autonomy.

Biodiversity in the Philosophy of Nature

From a Biological Approach

From a biological perspective, it is important to realize that the production of diversity is an intrinsic value of biological life. In Darwinian terms, living beings need to warrant their adaptation and that of their offspring within a variable environment by the optimization of their fitness. The best way to do so is to develop the ability to produce a diversity of adaptable phenotypes. When a living being produces a variation it is likely that the new characteristic will give advantage to its bearer, otherwise it is eventually removed by natural selection. Through the emerging genetic diversity, the differentiation of beings presents a better range of natural characteristics able to resist a maximum level of aggression from the environment, especially against diseases. Biodiversity is thus seen as an intrinsic value and source of goodness for biological life, because without it the very existence of life is in jeopardy.

From this it is possible to understand that biodiversity must not be reduced to the sum of different living beings in the biosphere, even the actual number of species that human beings should preserve and stop destroying. Virginie Maris reminds us that biodiversity is as much the outcome of a process as the process itself and all its potentialities (Maris 2010, 67). Biodiversity is effectively the capacity of life to produce new and different beings according to the environment and the state of evolution of life on Earth.

From Classical Philosophy

Classical philosophy also has its viewpoint regarding the intrinsic goodness of the diversity of natural beings. If the world is multiple, each different being, in its uniqueness, is an expression of its unique principle and origin, as the number one is the principle of all other numbers. In the *Timaeus*, Plato claims that the cosmos needs to be one in order to reflect the perfection of the divine, and then many, because divine perfection, oneness and simplicity need to be expressed by the greatest number of naturally existing entities. Each one thus bears an aspect of divine perfection (31a–33c). This idea will be re-expressed in the Plotinian reflection on the link between the One and the plurality of its expression through the diversity of natural beings (*Enneads, V* 3, 7, 8 13; *VI* 7, 15).

Consequences in Theology

This philosophical approach has had great fecundity in theological reflection through the ages, especially in the medieval theologies of Saint Bonaventure and Saint Thomas Aquinas. For both, the divine creative act produces true alterity, meaning something different from the divine nature, that is also a true reflection of God's being, especially its goodness and perfection. Therefore God's goodness and perfection are best expressed and reflected through a multiplicity of creatures.[4] Here we can see a part of what is expressed in Genesis. 1, when God sees that what he created was good.

In the relationship between the whole and the part, Aquinas says that creation has been created as a whole organized out of different elements that find their place according to their end in the system they form. For Aquinas there is more value in the whole than in its parts taken separately. This means that if one part is overly optimized in a system it is against the good of the whole. The various creatures therefore participate within creation according to their capacities for the good of the whole of creation, which is a better reflection of God's goodness (Aquinas, *Summa Contra Gentiles III,* 20, 1). Creation as a whole reflects God's goodness better than any individual creature. But to do that, creation needs all of these creatures.

Here we arrive at the idea according to which diversity, and subsequently the diversity of creatures, has an intrinsic value in God's perspective, which suffices to say that protection of biodiversity is a Christian value. Patriarch Bartholomew puts it thus:

> For human beings … to destroy the biological diversity of God's creation; for human beings to degrade the integrity of the earth by causing changes in its climate, by stripping the earth of its natural forests or destroying its wetlands; for human beings to contaminate the earth's waters, its land, its air, and its life – these are sins. (*Address*, quoted by Pope Francis in LS 8 or Patriarch Bartholomew I 1997, 216, quoted by Pope Francis in LS 8)

As we can see in these words, for two Christian world leaders such as Patriarch Bartholomew I and Pope Francis, biodiversity therefore has great ethical importance within Christian faith. I think we can go further in reflecting on the theological and ethical meaning of biodiversity. In the following section my claim is therefore that such diversity, including biodiversity, is a characteristic effect of continuous creation.

4. See St Bonaventure, *Breviloquium*, II, 3, 2; St Thomas Aquinas *Summa Theologiae, Ia* qu. 22, a.1, *resp.;* q. 103, a. 2, *ad* 3; q. 105, a. 5, *resp.*; *Summa Contra Gentiles*, II, 42, 1-2.

Continuous Creation and Biodiversity

What is Continuous Creation?

The meaning of this concept as it is understood here is taken from neither the Scholastic (see Suarez 2002, 112; Revol 2017, 17–58) nor Cartesian traditions (see Descartes 1967, 450; Revol 2017, 59–116). For the former, continuous creation is synonymous with the conservation of creation through time. God keeps creatures and the whole of creation in being through a permanent act of sustenance. This is in fact a human perception of God's creative act because of the temporal condition of our existence, but from the divine perspective, creation and conservation are one and a unique act. For the latter, continuous creation is the infinite repetition of creation ex nihilo at each moment of the passage of time. Here too, creation and conservation are understood in the same action. These two streams of thought however insist on the permanence of creation, not on the appearance of new beings. There is however a third approach to continuous creation that takes novelty into account (see Bergson 2001; Revol 2017, 141–219). It comes from the impact of the theory of evolution on the philosophy of nature. In this field continuous creation is revisited to interpret the natural creativity of the living through the successions of species in terms of divine immanent activity.

Heir to this latest stream of thought, the concept of continuous creation is nowadays widely used in the context of reflection on the dialogue between science and religion (Russell 1995, 10; Revol 2015, 29–116). It is a modern idea developed in the field of science/religion dialogue that claims that divine creative power is involved within time and temporality. But far from the arguments issued from the intelligent design theory implying a kind of supernaturalism (see for instance Dembski 2006, 723–5), it claims that natural evolutionary novelty – a basic element of process of differentiation process within living matter – is the outcome of a creative act that involves both creator and creature without any rupture in evolution's natural functioning, without any external interference by God's action within creation, but with God truly being the immanent sustainer of natural processes (Revol 2013, 702–11).

Continuous creation is not the running of a deterministic programme of the divine creative project. It is a theological interpretation of evolution as a creative process, incorporating the role of chance. Continuous creation is creation out of what precedes, and the latter is contingent. As chance is present at the different levels of natural selection, from mutation to the fitness of an organism in an environment, continuous creation gives theological meaning to the appearance of novelty in a natural context. As David Bartholomew puts it, chance is the most efficient means of divine creative – even purposeful – activity (Bartholomew 2008, 174).[5] Amidst the necessary structures of the universe, there are spaces of uncertainty allowing for an intrinsic creativity of creatures bringing forth new forms of being. As Arthur Peacocke said, there are propensities in the universe

5. In total opposition to intelligent design theory that exclude pure chance from its scope.

that indicate that all things cannot happen in this world, which is therefore not a chaos (Peacocke, 1993, 63–5). Contrary to the principle of plenitude that asserts that all the ideal possibilities of the universe must be expressed (Lovejoy 1936, 99–182), propensities are a signal that shows that only some of them can be contingently expressed according to the role of the constraints of natural laws and of environmental conditions.[6] But to speak of such propensities, such as the accumulation of information in memory among living beings, or such as complexity in more Teilhardian vocabulary, is an appropriate way of indicating that there are creative directions followed by creatures, according to the natural processes described by the sciences (Teilhard de Chardin 1955, 185).

God is participative in this process insofar as he empowers natural entities to do so, and he gives being to the novelty produced by these entities. Continuous creation therefore indicates a process of close partnership between God and creation as a whole, and in the case of biodiversity, with living beings in the process of evolution. This is then a process that requires a huge amount of time to be both efficient and then complete. God's participation as the very life of this process stresses an aspect of continuous creation which is important for the ethical perspective: God's omnipresence within creation. Even though this is a very classical aspect of theology and of God's attributes, it is also connected to the more recent emphasis put on God as immanent within the discussion on panentheism (Morry 1967, 699; Michon 2007, 991–3; Clayton and Peacocke 2004). Even if this topic is substantially developed in process theology (Cobb 2008), it is also a way of giving more meaning to this deep intuition developed by Aubrey Moore about evolutionary theory: 'The one absolutely impossible conception of God, in the present day, is that which represents Him as an occasional Visitor. … Either God is everywhere present in nature or he is nowhere' (Moore [1889] 1891, 73). Continuous creation strongly emphasizes God's presence in the creative dialogue constitutive of the process.

Furthermore, a reflection based on two areas of biblical and dogmatic tradition can help us to identify a Trinitarian dimension in this divine activity. For the purpose of our discussion we can focus on the Pneumatological aspect of continuous creation. According to Ps. 104.29-30, the breath of God is given to creatures so that they can live and be created; if God takes it back, creatures simply die. God's Spirit can thus be interpreted as a life-giver, as the creed of Constantinople puts it: 'the Lord and Giver of life'. Not unlike the role of the *élan vital* in Bergson's approach, the Holy Spirit can therefore be understood as the very life of the natural processes, especially those bringing forth novelty (Bergson 2001, 232). Thomas Aquinas already said that the third person of the Trinity, because he is the principle of God's goodness and generosity, is also the principle of the differentiation of creaturely forms (Emery 1995, 215). The Holy Spirit can thus be seen as a main actor of continuous creation. He is implied in the production

6. And if those conditions are gathered elsewhere than on Earth it might be then possible to think that life happened many times on other planets in our universe.

of new beings, and as such, he can be seen as the original fount of biodiversity, insofar as it is a part of the various kinds of diverse creatures existing in creation.

Continuous creation is thus a fitting concept for expressing the dynamic aspect of nature, read and theologically interpreted through the teachings of the story of life on earth, according to the framework of the theory of evolution.[7] Starting from the concept of continuous creation, it is possible to shape new ethical principles and landmarks indicating the intrinsic value of biodiversity. My claim here is thus that biodiversity, as a locus of God's creative activity, is a place of meaning and of the goodness of creation, and therefore a place that challenges current human behaviour towards nature and in this way advances behaviour attuned to the divine project of creation. This means that biodiversity is both a means and an end within the creator's project. It is a means insofar as it is a process of production of diverse forms. It is an end because as diversity it accomplishes its mission of creaturely features within the created world: reflecting God's perfection.

A Christian Vision of the Human Place in Creation

Amidst this profusion of quasi-infinite forms, it is legitimate to wonder, as Blaise Pascal did, about the loss of the place of the human being, caught between the different scales of infinity (Pascal 1998, 608–14). From the Scriptures, Christian tradition claims a special place for humankind within this universe. It is a particular creature, and the only one desired as such, as the pastoral constitution of Vatican II, *Gaudium et Spes*, puts it (§24). This means that in the midst of contingency, there is one creature that was awaited by God, and that is the human being (Bonaventure II, 4, 3-5).

However, placed within the scope of God's Incarnation in Jesus Christ, it seems clear that the human creature is necessary in order for there to be a being that can be said to be *capax Dei*, designating on the one hand a receiver of God's Revelation, and on the other a creature able to respond to God's initiative and thus able to receive God himself (see Moltmann 1981, 114–18). God's plan was to share the creaturely state and for that he needed a special creature. Here is an expression of Duns Scotus' tradition that says that original sin was not necessary for God to become incarnate. Incarnation was a plan of communion of the divine life shared between God and a created alterity able to communicate and respond to him (Delio 2003, 34).

It therefore seems that the meaning of biodiversity within creation is both the production of a maximum number of forms and the possibility of the creature *capax Dei* among them. The end of the process of continuous creation is thus clearly God's incarnation, as put by Arthur Peacocke: 'The incarnation in Jesus the Christ may then properly be said to be the consummation of the creative and creating evolutionary process' (Peacocke 2004, 216).

7. For a consistent philosophy of nature articulating theory of evolution with continuous creation see my work on *La nouveauté dans l'histoire de la nature*.

Interaction, a Means of Continuous Creation

The process of the production of biodiversity implies the concept of interaction of interdependence as constitutive interaction at different levels of intelligibility. On the biological level, ecology teaches us that a being is the emergence of an entity thanks to interactions it has with its natural environment (Miller and Ricklefs 2005, xx). Natural diversity is then the fruit of a complexity of interactions.

On the metaphysical level, the novelty that arises is grounded in the expression of a new substantial form that is created by the organization, arrangement and interaction of previous forms. The natural interactions are the expression of metaphysical interactions (Revol 2013, 804–24). Such interactions are possible because the forms selected for expression stem from a process among the interactions between God the Creator and creatures (885–992).

Creation is traditionally understood within the category of relationship, as Aquinas describes:

> The creature is the term of creation as signifying a change, but is the subject of creation, taken as a real relation, and is prior to it in being, as the subject is to the accident. Nevertheless creation has a certain aspect of priority on the part of the object to which it is directed, which is the beginning of the creature. Nor is it necessary that as long as the creature is it should be created; because creation imports a relation of the creature to the Creator, with a certain newness or beginning. (*Summa* q.45, a.3, ad. 2)

It means that creation as an act is a permanent connection between the Creator and the creature, the creation as a whole. Here it is an asymmetrical relationship because it is all in God's power to bring things into beings. It can be further understood under the analogy of the ecological relationship, according to Jürgen Moltmann, though in different terms and backgrounds. The creative Spirit inhabits the creation as his ecosystem (Moltmann 1997, 211–2). But the Holy Spirit does not need this creation ecosystem to be what he is. It is rather the place where he dwells, and by this dwelling it is the habitat itself which is enlivened.

However, what is interesting here is the idea of interaction that characterizes the act of creation as well as the production and creation of biodiversity. It seems that we have here a new expression of what Saint Bonaventure called the *vestigiae trinitatis*, the traces of God in creation (*Breviloquium*, II, 12, 1). Going further down this path, interaction is an expression of God's being Trinity. For Denis Edwards, continuous creation, and consequently its expression as biodiversity through the gathering of ecological relationships, can be understood as a signature or a footprint of God in creation (Edwards 2004, 205).[8] Pope Francis

8. 'Of course there is an infinite difference between created being-in-relation and the divine communion; but what continuous creation means is that created being-in-relation always springs from, depends upon, and in a creaturely way participates in, the being of divine Persons-in-Relation' (Edwards 1998, 389).

even puts thus it in *Laudato si*: 'For Christians, believing in one God who is Trinitarian communion suggests that the Trinity has left its mark on all creation' (§239). The fact that all is connected in the universe, in ecosystems, is a given of creation corresponding to both God's creative will and his Trinitarian nature (see also Revol 2016). I claim that this Trinitarian signature in creation carries strong ethical implications as it then shows that the denial of the interdependence in nature or even in human life is the source of many disturbances and forms of ecological destruction. This aspect of creation should be taken as a criterion for human action: showing respect for networks of interaction in nature in order to build a good life together in a world understood as a communion of relationships.

Furthermore, all creatures, with human beings among them, can truly be called co-creators when they bring true novelty into the world.[9] This has ethical consequences for human behaviour and activity, both in the field of human invention and in terms of human impact on nature. A way of life that tends to destroy biodiversity can be therefore understood as a counter-movement against continuous creation and against the Creator's project.

Towards a Philosophical Ethics of Ecology

It thus becomes important to situate the methodology of Christian ethics: Its proper purpose is that it should not be founded on pure religious discourse, but on an accessible rational discourse, meaning, and the search for universal and philosophical criteria. The reason for this approach is particular to its Christian starting point: because of the Incarnation of Christ, every aspect of human life concerns Christian discourse. Conversely, it also seems clear that every aspect of Christ's features concerns the whole of humanity. Thus there is an expectation, indeed a requirement, of a meeting and proper fit between Christianity and all that is human. Christian ethics must therefore first be trustworthy at the philosophical level; it must at least be translatable into philosophical speech attainable by natural reason. In this way the theological perspective may also provide a framework for secular philosophy.

The concept of continuous creation indicates that interaction, as a kind of which relationship is a species, should be at the heart of an ethics bringing forth the idea that true and real relationship is creative; it brings forth the flourishing of beings and the goodness of diversity.

Theology and Ethics

From this starting point we can summarize the theological aspects of biodiversity and their ethical consequences. First, as biodiversity is the totality of the living

9. A concept introduced by Claude Tresmontant in 1953 in the field of continuous creation (Tresmontant 1953, 37), then later with Adolph Gesché (1991, 153–84), and the theology of the Created Co-Creator of Philip Hefner (23-51). See also Peacocke (1998, 329–56).

creatures, it reflects God's glory better when the forms of living matter are many. Destroying it is akin to reducing this reflection. It is considered a sin by both the Orthodox Ecumenical Patriarch and the Catholic Pope. Second, as biodiversity is also a process of the production of diversity, it can be interpreted as the natural outcome of the divine process of continuous creation. The destruction of ecosystems and the humanly caused disappearance of species can be understood as human opposition to continuous creation and to the project of the creator. Third, in the midst of the production of novelty through the process of continuous creation humankind can share in production through its particular features: through free will and ingenuity humanity can also be a genuine source of novelty in the world, thanks to art, technique, politics and economics. And the more an outcome of human activity promotes true novelty, the more ethical value it bears. However, not all outcomes of human activity are necessarily genuine novelty, and this therefore needs to be explored.

Christian and Secular Ethics

There is something specific in the concept of continuous creation that can fit the needs of meeting the universal requirements of human reason. Participation in this process of continuous creation can be articulated in a naturalistic perspective on the basis of the concept of constitutive interaction found in ecology. What is really creative in nature is ecological interaction. It is the proper natural manifestation and even the seal of the divine creative act. It is what brings forth biodiversity and all its intrinsic goodness. This is a place where all understandings can meet, whether believing or non-believing. First because the meaning of diversity via reflection on the question of the one and the many is as old as philosophy, and second because we understand that what is vital is the role of life's creativity through biodiversity for the sustaining of living beings on earth.

Towards New Principles for Ecological Ethics

Constitutive interactions are creative because they are at the origin of the diverse forms of natural novelty that built the story of life on earth and the actual and evolving landscape of current biodiversity. The issue is: What is genuine novelty in the natural world, especially in living matter on earth? I propose to compile five criteria most able to specify the limits defining what is something new in nature.

First, novelty is contingent and unpredictable in the history of life. Second, novelty is the outcome of the set of dynamic interactions forming a system. Third, novelty must be finalized within the system. Fourth, novelty must be something sustainable, meaning that its structure is stable though time. Fifth, novelty must be the outcome of the process of integration. Integration here means the synergistic operation of several elements working together. If all these criteria are gathered so there is novelty, so far for the outcome of human activity which can then be considered as truly creative.

Creativity and diversity should thus be accepted as new universal principles for ethical reflection, alongside with vulnerability and autonomy, two other characteristics of true novelty that are also appropriate for describing biodiversity. Creativity and diversity are closely linked, because the second is an outcome of the first. Creativity, as a potentiality of nature, must be respected in order to guarantee the sustainability of life on earth through diversity that gives to living beings the means of adaptation and resistance to forms of external aggression.

Vulnerability is then derived from this reality, because every form of true natural novelty must prove itself through the screening action of natural selection before being confirmed as a natural and structural element of an ecosystem, and all life on earth is threatened by death and entropy. Autonomy eventually arises because it fits into the definition of true natural novelty as being something autonomous via the working of its constitutive interactions. It indeed brings into being an emergent entity or system that has irreducible complexity and an action on itself and on its environment (see Prigogine 1996; Kaufmann 1993, Sartenaer 2010, 371–404; Revol 2015, 239–51).

It is important to notice that what makes the unity and integration of these four principles, and gives full meaning to them, is the notion of constitutive interaction. I think that all actions and initiatives that may impact the natural environment should therefore reflect on these four questions: Does my initiative protect the potency of nature for producing forms of novelty? Does my action preserve and guarantee the safety of biological life by enhancing its possible diversity? Does my action favour the autonomy of a natural milieu or does it necessitate continual human intervention for its conservation? Does my action consider the vulnerability of life, ecosystem or species as worth protecting? And finally there is a fifth question underlying and subsuming the previous four: Does my action put systems, natural and human, in creative relationships?

Conclusion

This reflection on the meaning of biodiversity leads me towards the idea that new universal ethical principles can be defined via a theological approach seeking the meaning of continuous creation. The collaborative action of continuous creation between the Creator and creation itself is at the origin of a creation process of forms of natural novelty in time. The description of this process offers new representations of creation and hence of nature. This is a process that provides compelling reasons for believers to act as good stewards of creation, as co-creators involved in the work of continuous creation, and fight against ecological crisis. But the autonomous position of the creaturely partner leads us to consider that philosophical reflection on ecological interaction is quite sufficient to give ground to the ethical values of biodiversity through the establishment of four principles deriving from the concept of novelty: diversity, creativity, autonomy and vulnerability. These are complementary to the main idea of community of beings found in Aldo Leopold's land ethics. I hope they might be effective for a special reflection on ecological ethics, and maybe in other fields of ethics as well.

References

Bartholomew, David J., *God, Chance and Purpose, Can God Have It Both Ways?* (New York: Cambridge University Press, 2008).

Bergson, Henri. *L'Évolution créatrice* (Paris: PUF, 1941, 2001).

Bourgeois Léon, *De l'humanité* (On Humanity), Corpus des Œuvres de Philosophie en Langue Française (Paris: Fayard, 1985).

Bourgeois Léon, *Solidarité, L'idée de solidarité et ses conséquences sociales* (Solidarity, The Idea of Solidarity and Its Social Consequences) (Paris: éditions Le bord de l'eau, 2008).

Clayton, Philip and Arthur Peacocke (eds), *In Whom We Live and Move and Have Our Being, Panentheistic Reflections on God's Presence in a Scientific World* (Grand Rapids and Cambridge: Eerdmans, 2004).

Cobb, John B. Jr., *Back to Darwin, A Richer Account of Evolution* (Grand Rapids, MI and Cambridge, UK: Eerdmans, 2008).

Delio, I., *A Franciscan View of Creation: Learning to Live in a Sacramental World* (*The Franciscan Heritage Series* 2) (St Bonaventure and New York: The Franciscan Institute, 2003).

Dembski, William A., 'In Defence of Intelligent Design', in *The Oxford Hand Book of Religion and Science*, ed. Philip Clayton and Zachary Simpson (Oxford: Oxford University Press, 2006), 723–5.

Descartes, René. *Méditations touchant la première philosophie dans lesquelles l'existence de Dieu et la distinction réelle entre l'âme et le corps de l'homme sont démontrées* (Meditations Touching on First Philosophy in which the Existence of God and the Real Distinction between the Human Soul and Body Are Demonstrated), in *Œuvres philosophiques,* vol. II 1638–1642, ed. Ferdinand Alquié (Paris: Garnier, 1967).

Edwards, Denis, 'Original Sin and Saving Grace in Evolutionary Context', in *Scientific Perspective on Divine Action*, vol. 3 *Evolutionary and Molecular Biology*, ed. Robert J. Russell, William R. Stoeger and Francisco J. Ayala (Vatican and Berkeley: Vatican Observatory Publications/Center for Theology and the Natural Sciences, 1998), 377–92.

Edwards, Denis, 'A Relational and Evolving Universe Unfolding Within the Dynamism of the Divine Communion', in *In Whom We Live and Move and Have Our Being, Panentheistic Reflections on God's Presence in a Scientific World*, ed. Philip Clayton and Arthur Peacocke (Grand Rapids and Cambridge: Eerdmans Publishing Company, 2004), 199–210.

Emery, Gilles, *La Trinité Créatrice, Trinité et création dans les commentaires aux Sentences de Thomas d'Aquin et de ses précurseurs Albert le Grand et Bonaventure* (The Creative Trinity: Trinity and Creation in the Commentaries on the Sentences by Thomas Aquinas and His Precursors Albert the Great and Bonaventura) (Paris: Vrin, 1995).

Gesché, Adolph, 'L'homme créé créateur' (Man Created as a Creator), *Revue Théologique de Louvain* 2, no. 22 (1991)

Hefner, Philip, *The Human Factor, Evolution, Culture, and Religion* (Minneapolis: Fortress Press, 1993).

Hessel, Dieter T., and Radford Ruether, Rosemary (eds), *Christianity and Ecology* (Cambridge, MA: Harvard University Press, 2000).

Kauffman, Stuart, *The Origin of Order, Self-Organization and Selection in Evolution* (New York and Oxford: Oxford University Press, 1993).

Léonard, André, *Le fondement de la morale, essai d'éthique philosophique générale* (The Foundation of Morality: An Essay in General Philosophical Ethics) (Paris: Cerf, 1991).

Leopold, Aldo, *A Sand County Almanac, & Other Writings on Conservation and Ecology* (New York and Oxford: Oxford University Press, 1949, 2001).

Lovejoy, Arthur, *The Great Chain of Being* (Cambridge, MA: Harvard University Press, 1936).

Maris, Virginie, *Philosophie de la biodiversité. Petite éthique pour une nature en péril* (Philosophy of Biodiversity. A Little Ethics for Nature in Danger) (Paris: Buchet/Chastel, 2010).

Michon, Cyrille, 'Omniprésence divine', in *Dictionnaire critique de théologie*, ed. Jean-Yves Lacoste (Critical dictionary of theology) (Paris: PUF, 2007).

Miller, Gary L., and E. Ricklefs Robert, *Écologie* (Ecology) (Brussels: De Boeck, 2005).

Moltmann, Jürgen, *The Trinity and the Kingdom: The Doctrine of God*, trans. Margaret Kohl (London: SCM Press, 1981).

Moltmann, Jürgen, *God in Creation: An Ecological Doctrine of Creation*, trans. Margaret Kohl (London: SCM Press LTD, 1997).

Moore, Aubrey, 'The Christian Doctrine of God', in *Lux Mundi*, ed. Charles Gore (London: Murray, [1889] 1891).

Morry, M. F., 'Omnipresence', in *New Catholic Encyclopedia* (New York, Saint Louis, San Francisco, Toronto, London and Sydney: McGraw-Hill Book Company, 1967).

Pascal, Blaise, *Les pensées, Œuvres complètes*, vol. II (Coll. Bibliothèque de la Pléiade 34), ed. Michel Le Guern (Paris: Gallimard, 1998).

Patriarch Bartholomew, I., 'Beauty and Nature', Homily at the fiftieth anniversary dedication of St. Barbara Greek Orthodox Church in Santa Barbara, California, 8 November 1997, *Cosmic Grace + Humble Prayer, The Ecological Cision of the Grenn Patriarch Bartholomew I*, ed. John Chryssavgis (Grand Rapids, Michigan and Cambridge, UK: W.B. Eerdmans Publishing Company, 2003), 215–17.

Peacocke, Arthur, *Theology for a Scientific Age, Being and Becoming - Natural, Divine and Human*, Enlarged Edition (Minneapolis: Fortress Press, 1993).

Peacocke, Arthur, 'Biological Evolution, a Clue to the Meaning of Nature', *Scientific Perspectives on Divine Action*, vol. 3, *Evolutionary and Molecular Biology*, ed. Robert J. Russell, William R. Stoeger and Francisco J. Ayala (Vatican and Berkeley: Vatican Observatory Publications, Center for Theology and the Natural Sciences, 1998).

Peacocke, Arthur, *Evolution, the Disguised Friend of Faith, Selected Essays* (Philadelphia and London: Templeton Foundation Press, 2004).

Plato. *Timaeus.*

Plotinus. *Enneads.*

Prigogine, Ilya, *La Fin des certitudes, Temps, Chaos et Lois de la Nature* (The End of Certainties: Time, Chaos and Laws of Nature) (Paris: Odile Jacob, 1996).

Rémi, Beau, 'Ecocentrisme', in *Dictionnaire de la pensée écologique*, ed. Dominique Bourg and Alain Papaux (Dictionary of Ecological Thought) (Paris: PUF, 2015).

Revol, Fabien, *Le concept de création continuée. Histoire, critique théologique et philosophique, essai de renouvellement dans le dialogue de la théologie avec les sciences de la nature par la médiation de la philosophie* (The Concept of Continuous Creation. History, Theological and Philosophical Critique: An Essay in Renewal of the Dialogue with the Natural Sciences through the Mediation of Philosophy). Thesis for the obtention of a Canonical Doctorate in Theologie and a Canonical Doctorate in Philosophy, directed by François Euvé (Centre Sèvres) and Emmanuel Gabellieri (Lyon Catholic University, 2013).

Revol, Fabien, *La nouveauté dans l'histoire de la nature, herméneutique philosophique de la créativité naturelle* (Novelty in the History of Nature: Philosophical Hermeneutics of Natural Creativity) (Paris and Lyon: Vrin/IIEE, 2015).

Revol, Fabien, 'Les vestiges de la Trinité reflets du créateur dans la création continuée', in *Controverses sur la création: Science, Philosophie, Théologie*, ed. Bertrand Souchard and Fabien Revol (Controversies Concerning Creation: Science, Philosophy, Theology) (Paris and Lyon: Vrin/IIEE, 2016).

Revol, Fabien, *Le concept de création continuée dans l'histoire de la pensée occidentale* (The Concept of Continuous Creation in the History of Western Thought) (Paris and Lyon: Vrin-IIEE, 2017).

Revol, Fabien, 'Théologie de la création continuée' (Theology of Continuous Creation), *Revue des sciences religieuses* 91, no. 2 (2017): 253–69.

Russell, Robert J., 'Introduction', in *Scientific Perspectives on Divine Action*, vol. 2 *Chaos and Complexity*, ed. Robert J. Russell, Nancey Murphy and Arthur Peacocke (Vatican and Berkeley: Vatican Observatory Publications/Center for Theology and the Natural Sciences, 1995).

Sartenaer, Olivier, 'Définir l'émergence' (Defining Emergence). *Revue des Questions Scientifiques* 181 (2010): 371–404.

Sertillanges, Antonin-Dalmace, *La Philosophie de S. Thomas d'Aquin* (The Philosophy of St Thomas Aquinas), vol. 1, nouvelle édition revue et augmentée (Paris: Aubier Montaigne, 1960).

St Bonaventure. *Breviloquium: Works of St Bonaventure vol. IX*, trans. by Dominic V. Monti O. F. M. (New York: The Franciscan Institute Publications, 2005).

St Thomas Aquinas, *Summa contra Gentiles. Books II: Creation.* (Notre Dame: University of Notre Dame Press, 1976).

St Thomas Aquinas, *Summa contra Gentiles. Books III: Providence* (Notre Dame: University of Notre Dame Press, 1991).

St Thomas Aquinas, *Summa Theologica Part I ('Prima Pars')* (Altenmünster: Jazzybee Verlag, [1274] 2012).

Suarez, Francisco, *On Creation, Conservation, and Concurrence, Metaphysical Disputations 20, 21 and 22*, ed. Alfred J. Freddoso (South Bend, IN: St. Augustine Press, 2002).

Teilhard de Chardin, Pierre, *Le Phénomène humain* (The Human Phenomenon) (Paris: Seuil, 1955).

Tresmontant, Claude, *Essai sur la pensée hébraïque* (Essay on Hebrew thought) (Paris: Cerf, 1953).

Vatican Council II, *Pastoral Constitution on the Church in the Modern World, Gaudium et Spes* (7 December 1965).

White, Lynn Jr., 'The Historical Roots of Our Ecological Crisis', *Science* 155 (1967): 1203–7.

Chapter 7

AESTHETICS AT THE INTERSECTION OF SCIENCE AND THEOLOGY

Knut-Willy Sæther
Volda University College

Introduction

Mountaineering is one of my favourite activities. Many years ago, I was on a climbing trip with friends in Norway. I was walking alone, a little way from the group on our way back to base camp. Daylight changed to twilight and I sat down near a small lake and looked around the nature and landscape, with an immediate striking experience of beauty at the amazing details of the glacier buttercup (*Ranunculus glacialis*), overwhelming mountains and the blue horizon. My experience stimulated strong feelings as the beauty of nature opened up for wonder and the feeling of being part of a greater whole. A deep moment in nature took place.

Both before and after this experience, I have visited the same area, but have never had such an all-encompassing aesthetic experience of nature. I have also visited places, mountains, and sceneries more famous for their beauty and aesthetic environment. For me, however, that particular moment became *the experience*.

The desire to explore the role that our aesthetic experiences in nature plays in our reflection on 'our common cosmos' is based on the observation that aesthetics has been neglected in the traditional science–theology debate.[1] I will examine how insights from aesthetics can contribute to developing new approaches to the traditional science–theology dialogue, and explore how trajectories towards ethics arise from such insights.

First, I will clarify what kind of intersection and interdisciplinary approach I am advocating in such a context. My next step is to map out the role of aesthetics in light of the traditional science–theology debate and clarify the understanding of aesthetic experience and nature by analysing three experiences of nature considered as aesthetic: beauty, the sublime and wonder. I will explore the relationship between these three experiences of nature, and emphasize wonder as a core experience,

1. The science–theology debate to which I refer is elaborated later in the chapter.

with particular relevance for the intersection of aesthetics, theology and science. One of my claims is that wonder in this context is a significant phenomenon for further development towards ethics. I explore this trajectory enriched by Sigurd Bergmann's aesth/ethics and the notion of awareness, as main pointers for an ethical outcome.

What Kind of Interdisciplinarity Intersection?

Joe Moran notes that 'interdisciplinary' has become a buzzword, and is used across many different academic subjects (Moran 2002, 1). Interdisciplinary approaches are commonly claimed, and with the use of slogans such as dialogue, interaction and mutual enrichment, we hope to express that our own fields should not be understood as closed and limited from other fields. The emphasis on interdisciplinary approaches might be a reaction against modernity's divisions of fields and single areas of studying. On the other hand, the increasing complexity within fields, cumulative development of knowledge, and even knowledge crossing traditional fields, show some of the challenges of interdisciplinary navigation.

More urgent than discussing interdisciplinarity as such, is to clarify how we understand the relationship *between* fields, which is important when claiming intersection. Aesthetics, theology and science have different languages, methods and objects. One of many differences between science and theology can be followed along the lines of the *explanatory* dimension in science, and the hermeneutical dimension of *understanding* in theology. In addition, we also find variety concerning approaches and methods *within* the fields. If insights from different fields are related to each other, some sort of symmetry should be required. Do we find symmetry between aesthetics, science and theology? I cannot develop this thoroughly here, but stress the problematic in the case of theology. This field can be approached as specific theological disciplines and as philosophical theology (a holistic interpretation of reality informed by different fields). Well aware that the former aspect of theology needs to be taken into account in the same way we deal with insights from science and aesthetics, my entrance for establishing an intersection is philosophical theology.

For that reason, my position corresponds with Niels Henrik Gregersen's reflections on how science and theology might be related. He claims that interdisciplinary dialogue takes place as a transdisciplinary interpretation of the methods and results from different fields and always involves philosophical questions (Gregersen 1998, 186). Thus, the dialogue between aesthetics, theology and science takes place on a meta-disciplinary level.[2] Such an approach implies challenges, as we can be misled to believe that insights from different fields are independent from each other. We can claim interactions in various ways which

2. Gregersen later developed different types of theology (see Gregersen 2008, 290–310; Eikrem 2011, 152–66).

make the interdisciplinary scene more complex and correspond better to our many-layered reality, as developed by Robert John Russell in his model of creative mutual interaction and Mikael Stenmark in his multidimensional model (Russell 2008, 20 ff.; Stenmark 2004).[3] My point is to emphasize *the transdisciplinary level*, not how the particular campuses of science, theology and aesthetics might inform and challenge each other in different ways. Such a transdisciplinary level does not privilege different fields, but is a philosophical–theological reflection informed by insights from aesthetics, science and theology. Thus, this chapter will later emphasize the route from aesthetics towards wonder on a transdisciplinary level.

Intersection might not appear as a field in itself, but is a dynamic and developing area for reflection where no one can claim full ownership, as pointed out by Holmes Rolston: 'The interface between science and religion is, in a certain sense, a no-man's land. No specialized science is competent here, nor does classical theology or academic philosophy really own this territory. This is an interdisciplinary zone where inquirers come from many fields. But this is a land where we increasingly must live' (Rolston 2006, viii). This is surely the case when we add aesthetics to this picture. Even if it is no-man's land, it is surely a land, strongly contextual, coloured by our cultural and social settings.

We should not insist on strong ties between the different fields, as Rolston elegantly has expressed it: 'The religion that is married to science today will be a widow tomorrow' (Rolston 2006, ix). However, he also says religion divorced from science today will leave no offspring tomorrow. What makes this both interesting and challenging is that insights from one field, such as aesthetics, might resonate with insights from other fields, such as science or theology. However, the transdisciplinary interpretation has to be explorative and, at the same time, constantly be aware of the problem of making tight connections between fields.

Aesthetics and the Science–Theology Dialogue

The assertion that aesthetics has been neglected in the science–theology debate is dependent on *which* science–theology debate we have in mind. My backdrop is the scholarly discussion in the Anglo-American world, starting with Ian G. Barbour's groundbreaking work in late 1960s. This dialogue affords only a marginal role for aesthetics, but at least one modification is needed, as beauty as phenomenon has been elaborated among some scholars in the science–theology dialogue. One example is the beauty of nature as a part of different types of revised natural theology (McGrath 2008; Polkinghorne 1998, 69 ff.).

One reason that aesthetics circulates in the periphery of science–theology might be found in the traditional self-understanding of aesthetics as philosophy of art (Stecker 2005, 1). A broader approach, however, partly influenced by recent

3. The idea of a many-layered reality in this context comes from in particular Arthur Peacocke and Michael Polanyi (see also Losch, this volume).

research in fields such as environmental aesthetics and theological aesthetics is becoming more prolific. Further, we find new approaches in which aesthetics is strongly emphasized as a constructive dialogue partner. These approaches, in combination with insights from other scholarly discussions such as writings in eco-theology and theological aesthetics, give us a broader understanding and deeper insight of the role of aesthetics relevant for the intersection of science and theology.

Aesthetics resonates in the natural sciences. I will here give two examples which emphasize the relationship between aesthetics and the natural sciences, related to beauty as phenomenon. The first concerns scientific language, the second concerns insights from biology. In science, nature is the object studied, and through the scientific language of mathematics, we gain an insight of nature as beautiful. The scientific enterprise presupposes among most scientists some sort of critical realism, which implies that it is possible to describe nature through mathematics alone, such as through formulating models and theories about the physical reality. In making scientific theories there are non-empirical criteria involved, however, and some of these criteria are aesthetical, such as elegance and simplicity. Einstein, Dirac and Weinberg all advocated such an approach (McAllister 1996, 34 ff. and 96 ff.).

My second example is related to biology. According to Alejandro García-Rivera, who brings beauty of nature to the core of the interdisciplinary, insight from evolutionary developmental biology, combined with theories of beauty from aesthetics, gives us trajectories for how we can more fully understand living dynamic forms in nature. Beauty is thus an objective property of natural systems, pointing at aesthetic qualities *in* nature, and not only in the eye of the beholder (García-Rivera, Graves and Neumann 2009, 243–63). Hence, aesthetic experiences of nature resonate in science. In the following section I turn to the field of aesthetics to make some necessary clarifications of aesthetic experiences and nature which are useful to further my analysis.

Aesthetic Experiences and Nature

Within aesthetics, the question of what an aesthetic experience actually is has been widely discussed and this illustrates the complexity of aesthetic experiences.[4] Some of the questions are: Does an aesthetic experience provide some insight? Is the experience active or passive? Is it contemplative? Is the experience unique compared to other experiences? Adele Tomlin claims that complexity is at the core of aesthetic experiences and this displays something genuine of what an aesthetic experience is. Therefore, we need to avoid modernistic reductions of complex wholes: 'Buddhist and Asian philosophies, existential phenomenology,

4. This is also the case for the more narrowed understanding of aesthetics as theory of art, for instance in Carrol (2006, 69 ff.) and Shustermann and Tomlin (2008, 1).

hermeneutics, deconstruction, postmodernism and philosophical pragmatism have seriously challenged the reduction of complex wholes to simple constituents and the hegemony of scientism in the field of cultural experience and knowledge' (Shusterman and Tomlin 2008, 2). Tomlin's point on the complexity of aesthetic experience is useful to have in mind when we turn to experiences of nature, which in recent years have been increasingly emphasized in the field of aesthetics (Bisgaard 2005, 6–13).

Arnold Berleant describes aesthetics of nature as a theory of sensibility (see also Stecker 2005, 28 ff.). This sensibility is a 'perceptual awareness that is developed, guided, and focused' (Berleant 2012). Berleant makes a distinction between simple perception and educated perception, emphasizing the latter, understood as knowledge-based, informed, perception. If we value nature in a certain way, we approach nature by having expectations of what might take place in nature; therefore, no experiences of nature are value-free since our encounter is partly shaped by our interests and what we pay attention to. Berleant continues: 'Thus the search for the satisfaction of sensible experience comes from a thirst for positive perceptual value, which we find can be fulfilled by the perceptual satisfactions of the arts, natural beauty, and the immensely rich perceptual forms and details of the world of human life' (Berleant 2012, 374–86). On the other hand, aesthetic experiences of nature are not necessarily expectation-dependent; they might be regarded as unexpected, striking and surprising. In both cases we appreciate the experiences.[5]

One common aspect of beauty, the sublime, and wonder in nature is their intense character as experiences. Martin Seel argues that aesthetic experience is an intensified form of aesthetic perception and that 'aesthetic perception is an attentiveness to the appearing of what is appearing' (Shusterman and Tomlin 2008, 7). An aesthetical experience necessarily differs from daily, regular, experiences, says Seel. In addition, according to Tomlin, aesthetic experiences have a *transformative* dimension (Shusterman and Tomlin 2008, 2).

Beauty, the sublime, and wonder can be considered as meeting places for aesthetics, theology and science. Even though they can be understood as aesthetical experiences, they are not limited to the field of aesthetics and ought to be developed in an interdisciplinary context.[6]

The term 'nature' is not problematized in depth here, as this has been done elsewhere (Proctor 2009). Here we can understand nature in broad terms, both

5. Within aesthetics, environmental aesthetics has emphasized the topic of aesthetic appreciation of nature (see Berleant 1998, 114–20).

6. A wide range of experiences are present in our encounter with nature widely understood. Two of them, not examined here, are the experience of nature's ugliness and meaninglessness. John Cottingham says nature is utterly blank, silent, impersonal power that will 'neither care nor know' (Cottingham 2014, 67). Thus, the various experiences of nature, including the example from Cottingham, have to be taken seriously as experiences, also in relation to other experiences of nature.

as bio-physical and as non-biophysical nature.[7] My use of the term 'nature' also has cosmological connotations, including nature understood as micro- and/or macrocosm. In addition, nature is interwoven with culture in different senses. There is barely an area of untouched nature remaining, and humans are both spectators, participants in, and part of nature (Bergmann 2011, 30; Rolston 2002, 130; Kemal and Gaskell 1993, 1 ff.).[8]

Beauty, the Sublime and Wonder

How are we to understand beauty, the sublime and wonder as experiences of nature? Ross Wilson puts it this way: 'The philosophical discipline dedicated to the investigation of beauty and sublimity in nature and art, [is] the discipline of aesthetics' (Wilson 2015, 419). I will add wonder as a related phenomenon to the other two. Jeffery G. Sobosan says the experience of gazing at the stars on a clear night sky evokes several feelings, and is both an aesthetic and a moral experience.[9] Such an experience is about wonder: 'What [the stars] stir in me was wonder, and I have turned this wonder loose in my imagination many times as I have looked into the clear night sky. And each time I have been given joy' (Sobosan 1999, 2). Thus, all three experiences are interwoven and can be understood as aesthetical experiences.

The experience of beauty of nature can be approached in different ways. One example is the theological cosmology developed by García-Rivera. He emphasizes the beauty of nature in living dynamic forms (García-Rivera 2009, 81 ff.). His reflections are based on insights from developmental evolutionary biology. As I have mentioned, he identifies aesthetical patterns in living dynamic forms. He says we need aesthetics to fully understand these forms. The insight from aesthetics for such a deeper understanding provides the traditional definitions of beauty: 'unity in variety' and 'freedom from internal contradictions'. On the other hand, García-Rivera lacks a broader understanding of beauty of nature. I will expand the understanding of beauty of nature in two ways.

First, we need to include beauty in our experiences of non-biological nature as well. Resources from aesthetics, such as 'unity in variety' and 'freedom from internal contradictions' can also fit into such a broader notion of beauty of nature. Second, a broader approach to beauty in nature can be developed on different scales. We notice this as an important perspective in the field of environmental

7. On different understandings and approaches to nature, see Proctor (2009, 3 ff.).

8. According to those such as Stecker, how we experience and appreciate nature is so different. We therefore need to problematize the relationship between the particular and holistic in such an experience (see Stecker 2005, 1).

9. Sobosan says: 'It is aesthetic because the panorama of shapes and color my eyes take in produce an experience of beauty; it is moral because knowledge of the sheer size and age of what I am seeing produces an experience of humility' (Sobosan 1999, 1).

aesthetics (Berleant 2005, 31). Thus, experience of beauty in nature can take place both on a smaller scale, such as a piece of rock or flower, and on a larger scale, such as landscapes and scenery (Griffin 2011, 13). The former is an approach resonating with Kant's emphasis on beauty of particular objects in nature. The latter, the beauty of landscapes, is an aesthetic experience as well, as pointed out by Alister McGrath: 'Landscapes often evoke a powerful aesthetic response in humans' (McGrath 2008, 270). From the field of aesthetics, Ronald W. Hepburn points out a similar view, and problematizes what we actually mean by aesthetic appreciation of nature. For Hepburn, all sizes and kinds of natural landscapes have to be included (Hepburn 1993, 65 ff.). The appreciation of nature includes a wide approach to the experience of beauty, as the rewarding-to-contemplation, a great range of emotional qualities, without necessarily being pleasurable or lovable or suggestive of some idea.[10]

My second experience of nature, the sublime, can be followed along the same trajectories as for beauty. However, the sublime is an aesthetic experience which differs from the experience of beauty, as pointed out by Kant and Burke (Zuckert 2012, 64). For Kant, the sublime is related to natural phenomena. Burke emphasizes that the sublime is a stronger emotional experience than beauty (Mothersill 2004, 155; Brady 2013, 22). He associates the sublime with the experience of fear and horror, and therefore understands the sublime as raw and dark, and not solely as an experience of grandeur.[11] According to Elaine Scarry, the sublime and beauty are understood as more separate phenomena through the Enlightenment, and the sublime was used to describe nature, while beauty was less emphasized (Scarry 1999, 85). Kathryn Alexander describes this change in the aesthetics of nature as the decline of beauty (Alexander 2014, 53).

Experiences of beauty and the sublime in nature are related (Jørgensen 2012, 37). However, they evoke different feelings. The beauty of a landscape, such as the scenery of mountains on a sunny day, may well evoke changes in our feelings in the direction to the sublime when the dark clouds are approaching, the wind is increasing, and it starts to rain. The deeper and more complex feelings evoked when we experience the sublime in nature is an indication of something reaching beyond natural limits. This can be related to transcendence in a specific way. McGrath describes transcendence as 'reaching beyond natural limits, an awareness of something that lies beyond the boundaries of human experience' (McGrath 2008, 42). This can be related to the sublime, which evokes something in the depth of humans, exceeding regular daily life. McGrath says that

10. By emphasizing beauty, we run the risk of over-romanticizing nature and leaving out other important experiences of nature. It is not possible for me to develop this further here, only to indicate that nature experienced as ugly, dreadful, raw and even meaningless, are some perspectives we need to keep in mind. The experience of the sublime in nature can thus be related to such a broader approach of experiences.

11. The sublime can therefore be related to wilderness as well (see Bohannon 2014; Griffin 2011, 13 ff.).

such an experience is 'an enlargement or expansion of the human spirit, linked with sensible objects, yet transcending them' (McGrath 2008, 42). The sublime and transcendence are interwoven, and we are experiencing the sublime in the borderland to transcendence, he says.

The third experience is wonder. This phenomenon is intertwined with beauty and the sublime, although it differs from the other two in several ways. Wonder is a broader term than beauty and sublime, including many perspectives which become easily blurred (Deane-Drummond 2006, 1).

In the context of this chapter, wonder as an experience of nature is understood as taking place together with, and in the extension of, the experiences of beauty and the sublime. When we experience beauty and the sublime in nature, this gives rise to wonder, as wonder can take place as a reflection evoked by what we perceive (Ledley 2009, 247). Derek Matravers characterizes this type of wonder as a reflective state. The experience of beauty and the sublime in nature evoke a first-order non-cognitive state in us and Matravers describes this feeling as astonishment (Matravers 2012, 168).[12]

Wonder can be understood at least in two directions as an experience of nature, by having both a passive and an active dimension. The former is when our perceptions and experiences of something in nature evoke the feeling of wonder. The passive dimension of wonder has similarities to beauty and the sublime, as wonder is evoked from the perception of nature. In one sense, nature hits us and we might talk about the experience of wonder, as we do with experience of the sublime and experience of beauty. However, wonder is a larger, more overreaching, concept, and, in Robert Fuller's words, described as one of the most profound and subtle of all human experiences (Fuller 2012, 64).

The latter, the active dimension, is when wonder functions as a motivation or inspiration for understanding. Hence, wonder is also about curiosity. This motivation is a kind of an inner flow, for instance, for the scientist as described by Ralph Waldo Emerson: 'Men love to wonder, and this is the seed of science' (quoted in Ledley 2009, 246). More broadly, we might describe this inner flow as a shared experience for all of us in our search for understanding of the world. Such a flow is not only intellectual, as the search for understanding. The active dimension of wonder can also be understood as motivation for practices. This is emphasized by Sophia Vasalou. Although she describes philosophy as the origin for such a practice, she points out that wonder as a practice is shared through the ages by other disciplines, including science (Vasalou 2012, 2). This is also the case

12. Maltravers points out three aspects of wonder with relevance for valuing wonder. The first one is this reflective state. The second is wonder as a state that has distinct affective and cognitive elements, where the burden of the value attached to it is carried by the latter. The third is wonder as an affective state, but of a complex and valuable kind. He argues that the best defence of the value of wonder is the third one, which take into account that our experience is affected by our beliefs, namely that our object is profound and impressive in relation to us. Thus, wonder is only valuable if that belief is true (see Maltravers, 177).

for religion, she says. Wonder plays a major role in religious practices: 'In religious and spiritual practices, again, wonder – along with its associated concepts, such as awe – has sometimes been seen as the religious passion *par excellence*' (Vasalou 2012, 2). Based on Vasalou's approach we might understand wonder both as natural wonder and practised wonder, and these are closely related. Whether we emphasize the passive or active dimension of wonder, they can be explored as experiences with relevance for ethics.

From Experiences of Nature towards Ethics

My approach to experiences of nature indicates a positive evaluation of them, claiming an appreciation of nature, which is a value judgement. When experiencing beauty and the sublime in nature, we are drawn towards something, which in a deep sense expresses our love and attachment to nature (Parsons 2008, 5).[13] This expands the Kantian disinterested contemplation. According to Martin Seel, the aesthetic of nature is an ethics of the individual conduct of life: 'For aesthetics, being concerned with specific forms of and opportunities for process-oriented activity, is generally part of an ethics of the good life' (Seel 1998, 342). Hence, aesthetics of nature and ethics are related.

We need to ask what the interdisciplinary benefit of this actually is. Aesthetic experiences of beauty, the sublime and wonder in nature have resonance in various fields, such as science and theology. As mentioned earlier, beauty can illustrate the cross-disciplinary character of these experiences, since scientists deal with beauty as given in nature and expressed through the language of mathematics (McAllister 1996). However, beauty is not an object per se for science, but a non-empirical criterion in constructing scientific theories (Sæther 2011, 46). Holmes Rolston argues: 'Science becomes the primary avenue for perceiving landscapes. Better than any other – necessary though not sufficient for their most adequate understanding' (Rolston 1995, 380). Scientific insight is not a precondition for aesthetic experience, but can enrich our experience. New knowledge stimulates for further wonder (Español-Echániz 2010, 49). For theology's part, beauty has been explored along the lines of a quality in God's creation, and God as beautiful and the origin of beauty, especially emphasized in the context of theological aesthetics.

In terms of wonder, we find potential for a rich platform for further exploration, related to the other two experiences. Despite wonder's many faces, wonder as experience of nature, evoked through and in the extension of beauty and the sublime, exhibits something stimulating for both our feelings and our cognition. Wonder can be thought of as an experience related in a specific way to the prior experiences of beauty and the sublime. Wonder as a core experience resonates

13. This does not exclude an approach of wonder based on other experiences, such as evil. However, this opens the discussion along different lines which I will not develop here.

in aesthetics, science and theology, gives a possible route towards ethics. Robert Fuller's reflections on wonder can be helpful in this respect. He says wonder expands 'our cognitive repertoire in ways that facilitate a stance toward life that might be considered broadly spiritual' (Fuller 2012, 67). Fuller refers to Dacher Keltner and Jonathan Haidt, who point out that evolution favoured a specific way of motivational programme described as moral emotions. These emotions differ from those directed towards physical safety. Wonder can be understood as one of these moral emotions, he says.

Fuller's next move is to explore how awe fits into this picture. Awe transforms humans' life, goals and values (Keltner and Haidt 1999, 505–21; 2003, 297–314). Awe evokes the feeling of being a part of a larger whole. Wonder is closely related to awe, says Fuller. Both are caused by novel and unexpected stimuli challenging already conceptual categories. However, wonder also differs from awe, as the latter is more directly accompanied by a certain fear or subordination. Based on the similarities between awe and wonder, Fuller claims that humans are able to experience wonder because the brain has developed through natural selection in such a way to respond on new stimuli, such as greatness and vastness. This creates a need for systematizing these experiences in a larger picture.

According to Fuller, wonder 'evokes the subjective sense that we have established a harmonious relationship with the widest possible range of human experience' (Fuller 2012, 81). Wonder is characterized by an unlimited openness, it slows down our active will and makes contemplation of something larger possible. Further, wonder is succeeded by happiness and a feeling of expanded horizons, which makes room for new possibilities and even creates novelty.

All these characteristics and outcomes of wonder make this phenomenon different from what we can describe as everyday experiences. For Fuller, human experiences are thought of in two ways: Either they are taken for granted in our everyday life (as profane experiences) or they exceed the profane in such a way by stimulating meaning and intention. Such experiences, filled with wonder, carry us behind everyday life, and contain in a deep way mystery. Fuller continues: 'They [experiences of wonder] enable us to view the world as it exists independently of our own immediate needs, and they thereby foster empathy and compassion' (Fuller 2012, 85). Hence, wonder has ethical implications.

Guided by Fuller and his interdisciplinary approach, we see a route from wonder towards ethics. If we add my remarks from the previous section on aesthetic experience, we find this resonating in light of wonder. Wonder as an aesthetic experience is complex, consisting of a perceptual awareness, and is both intensifying and transformative. What is not articulated fully in Fuller's account with relevance for bridging wonder and ethics is the notion of awareness, which may be understood as part of what Fuller describes as motivational programme. In the coming sections I turn to Bergmann's concept of aesth/ethics and the notion of awareness. This can contribute to a more profound understanding of how experiences of nature (beauty, the sublime and wonder) can be related to ethics.

Bergmann's Aesth/ethics and the Notion of Atmosphere

Bergmann coined the term 'aesth/ethics' to emphasize aesthetics in the forefront of both theology and ethics (Bergmann 2011, 31 ff.). By this, he creates a creative interdisciplinary field, including topics such as eco-theology, lived space, pneumatology and liberation. Aesthetics is understood in the broad sense as a discursive and artistic production and a reflection of practices and discourses on what he describes as synaesthetic perception, creation and reception. Bergmann shows that aesth/ethics (with the slash) indicates that ethics is embedded continuously in perception. He says if ethics is defined as a discursive reflection on moral problems, we cannot exclude people's mental capacities and separate aesthetic competence from moral competence, thus the perception of moral problems must be prior to their reflection and possible solution (Bergmann 2011, 31). The slash is intended to avoid privileging ethics. By perceiving oneself and others in a common environment will not only precede but also regulate continuously moral agency and reflection. For Bergmann this concept is further developed theologically where he emphasizes pneumatology as an ecological aesth/ethics of the Spirit where the spatial dimension is crucial.

Bergmann develops three concepts for an eco-theological aesth/ethics of the Spirit in lived space: inhabitation, *Beheimatung* and atmosphere. Inhabitation challenges theologians to emphasise the perception of space and life. The pneumatological lens gives a focus on practices with and discourses about the 'synergy of the Holy Spirit inhabitation and cohabitation built and other environments' (Bergmann 2011, 33). *Beheimatung*, the German word for making oneself at home, is the second concept. The question we need to ask is how to make oneself at home at 'Earth, our home' which we ourselves are spoiling, asks Bergmann.

The notion of atmosphere is used to dissolve the Cartesian dualism; the distinction of subject and object. Bergmann draws on resources from phenomenology to allow a theological interpretation of the Spirit dwelling among us in different interspaces and scales of creation. The important question to ask, says Bergmann, is how and where the Spirit is at work. The message in the Christian doctrine of the Spirit is to say that the Holy Spirit can work in, with and through all places, spaces and scales of creation (Bergmann 2011, 34). Such a pneumatological approach can be problematized, but at least it opens up the potential for an understanding of the Spirit as not bound by human's limitations. Thus, atmosphere is God's traces on earth. For Bergmann, atmosphere intertwines us with nature in a profound way, which sheds light on what aesthetics is all about. Referring to Gernot Böhme, Bergmann says aesthetics can be understood as a self-aware human reflection on one's living-in-particular-surroundings (Bergmann 2006, 336). For Bergmann this expresses an ecological aesthetics where humans should participate in the natural life cycle. Atmosphere is a keyword in this aesthetics, emphasizing interconnectedness of the inner and outer, the bodily and the spiritual, the surrounding and the inhabitation.

This has impact for our approach to ethics, as aesthetic competence and moral competence are intertwined. Bergmann says the perception of moral problems must be prior to their reflection and solution: 'It takes both a sharp mind *and* the capacity of the senses to see our neighbor's misery, to answer Cain's question: "Lord am I my brother's keeper?"' (Bergmann 2006, 337). Underlying Bergmann's move from aesthetics to ethics (or more rightly expressed as how aesthetics and ethics are woven together) is the notion of awareness. This can be helpful for a further development of my trajectory from wonder towards ethics.

Awareness: Bridging Wonder and Ethics

According to Bergmann and Eaton, the term 'awareness' in the Western tradition has its roots in Greek, indicating the ongoing reflection on oneself in the midst of disturbing worldly attractions (Bergmann and Eaton 2011, 3). Awareness was a self-awareness, understood as a practice that was achieved by 'wakefulness, attentive observation and regular introspection' (Bergmann and Eaton 2011, 3). From this, the notion of awareness developed in different contexts, such as theology and modern phenomenology. A lack of awareness creates alienation, and awareness is a skill to be nourished and developed.

Awareness is an aspect of *how* we sense and perceive the world in a specific way. The way of seeing things is prior to the way of acting, it is about our senses and perceptions and how we actually pay attention. We might have primarily in mind the visual senses when we address awareness, but all our senses are at work; what we see, taste, hear and touch.

The term 'awareness' can be understood in different ways, and yet, in related directions, such as consciousness, knowledge and understanding. Other aspects of awareness are openness and receptiveness. Bergmann and Eaton say: 'What we become aware of influences the kinds of questions we ask, how and what we reflect upon, and ultimately how we answer our queries' (Bergmann and Eaton 2011, 3). To be aware not only is something we start to do in a specific context, but is also about staying aware, according to Bergmann and Eaton. They present three main approaches to awareness in their ecological context: (1) Awareness of Earth as a lived space, (2) Awareness of climate change (3) Awareness of our common future (Bergmann and Eaton 2011, 4–5). We see clearly that these 'approaches to awareness' have strong ethical relevance.[14]

Bergmann and Eaton specifically point out the relationship between awareness and ethics. On the other hand, what is the relationship between wonder and awareness? Within the context of Judaism, Abraham Heschel says: 'Awareness of

14. Similar thoughts are developed in another context by Andreas Losch: 'Human beings can *cooperate* in creation, when they are aware of their own creator, which incorporates awareness of being bound to nature' (2005, 285).

the divine begins with wonder. It is the result of what man does with his higher incomprehension. … Wonder or radical amazement, the state of maladjustment to words and notions, is therefore a prerequisite for an authentic awareness of that which is' (Heschel 1955, 46). We can reformulate Heschel's notion of awareness slightly by saying awareness of nature begins with the experience of wonder. Wonder is a prerequisite for an authentic awareness of that which is. Such an understanding weaves awareness together with wonder, and then contributes to an attempt to bridging wonder and ethics.

Conclusions

Aesthetic experiences of nature is complex. This is the case for all three experiences of nature explored in this chapter. In particular, this is the case for wonder, which is evoked by the experiences of beauty and the sublime. Wonder has an intensified and transformative character, is a phenomenon expanding our needs, and fosters empathy and compassion.

Bergmann's notion of aesth/ethics shows one approach to how intertwined aesthetics and ethics are. Aesthetics as self-awareness on one's living-in-particular-surroundings gives trajectories for exploring aesth/ethics eco-theologically along the lines of inhabitation, *Beheimatung*, and atmosphere. Underlying this move from wonder to ethics is the notion of awareness. Awareness begins with wonder and emphasizes how we sense the world, and the way we sense influence how we act.

Thus, wonder, awareness and an intertwined understanding of aesthetic and ethics creates a common stream towards ethics. How this might be developed further is in the extension of my analysis, although some possible routes are developed in eco-theology. For the future, we need a science–theology dialogue, enriched by aesthetics, which to a greater degree emphasizes ethical concerns, and thus helps us to open ourselves for other creatures, for the whole creation and our common cosmos.

Reference

Alexander, Kathryn B., *Saving Beauty: A Theological Aesthetics of Nature* (Minneapolis: Fortress Press, 2014).

Bergmann, Sigurd, 'Atmospheres of Synergy: Towards an Eco-theological Aesth/ethics of Space', *Ecotheology* 11, no. 3 (2006): 326–56.

Bergmann, Sigurd, 'Aware of the Spirit: In the Lens of a Trinitarian Aesth/Ethics of Lived Space', in *Ecological Awareness: Exploring Religion, Ethics and Aesthetics*, ed. Sigurd Bergmann and Heather Eaton (Münster: LiT Verlag, 2011), 23–39.

Bergmann, Sigurd, and Heater Eaton, 'Awareness Matters', in *Ecological Awareness: Exploring Religion, Ethics and Aesthetics*, ed. Sigurd Bergmann and Heather Eaton (Münster: LiT Verlag, 2011), 1–7.

Berleant, Arnold, 'Environmental Aesthetics', in *Encyclopedia of Aesthetics vol. 2*, ed. Michael Kelly (Oxford: Oxford University Press, 1998), 114–20.

Berleant, Arnold, *Aesthetics and Environment: Variations on a Theme* (Aldershot: Ashgate, 2005).

Berleant, Arnold, 'Aesthetic Sensibility', *Ambiances* (2012). https://ambiances.revues.org/526

Bisgaard, Ulrik, 'The Return of the Aesthetic Experience of Nature – Historical and Present Conceptions', *Nordic Journal of Aesthetics* 17, no. 32 (2005): 6–13.

Bohannon, Richard, *Religions and Environments: A Reader in Religion, Nature and Ecology* (London: Bloomsbury Academic, 2014).

Brady, Emily, *The Sublime in Modern Philosophy: Aesthetics, Ethics, and Nature* (Cambridge: Cambridge University Press, 2013).

Carroll, Noël, 'Aesthetic Experience: A Question of Content', in *Contemporary Debates in Aesthetics and the Philosophy of Art*, ed. Matthew Kieran (Malden, Oxford and Carlton: Blackwell Publishing, 2006).

Cottingham, John, *Philosophy of Religion: Towards a More Human Approach* (New York: Cambridge University Press, 2014).

Deane-Drummond, Celia, *Wonder and Wisdom: Conversations in Science, Spirituality and Theology* (London: Darton, Longman and Todd Ltd., 2006).

Eikrem, Asle, 'Dogmatikk som samtidsteologi. En kritisk videreutvikling av N.H. Gregersens program', *Dansk Teologisk Tidsskrift* 74 (2011): 152–66.

Español-Echániz, Ignacio, 'Aesthetic Experience of (landscape) Nature', *Enrahonar* 45 (2010): 41–50.

Fuller, Robert, 'From Biology to Spirituality: The Emotional Dynamics of Wonder', in *Practices of Wonder: Cross-disciplinary Perspectives*, ed. Sophia Vasalou (Eugene: Pickwick Publications 2012), 64–87.

García-Rivera, Alejandro, *The Garden of God: A Theological Cosmology* (Minneapolis: Fortress Press, 2009).

García-Rivera, Alejandro, Mark Graves and Carl Neumann, 'Beauty in the Living World', *Zygon* 44, no. 2 (2009): 243–63.

Gregersen, Niels Henrik, 'A Contextual Coherence Theory for the Science-Theology Dialogue', in *Rethinking Theology and Science*, ed. Niels Henrik Gregersen and J. Wentzel van Huyssteen (Grand Rapids and Cambridge: W. B. Eerdmans Publishing Company, 1998), 181–231.

Gregersen, Niels Henrik, 'Dogmatik som samtidsteologi', *Dansk Teologisk Tidsskrift* 71 (2008): 290–310.

Griffin, John, *On the Origin of Beauty: Ecophilosophy in the Light of Traditional Wisdom* (Bloomington: World Wisdom Inc., 2011).

Hepburn, Ronald W., 'Trivial and Serious in Aesthetic Appreciation of Nature', in *Landscape, Natural Beauty and the Arts*, ed. Salim Kemal and Ivan Gaskell (Cambridge: Cambridge University Press, 1993), 65–80.

Heschel, Abraham Joshua, *God in Search of Man: A Philosophy of Judaism* (New York: Farrar, Straus and Giroux, 1955).

Jørgensen, Dorthe, 'Fornemmelsens filosofi. Æstetik, fænomenologi og erfaringsmetafysik', in *Filosofi og kunst*, ed. Ulla Thøgersen and Bjarne Troelsen (Aalborg: Aalborg Universitetsforlag, 2012), 33–45.

Keltner, Dacher, and Jonathan Haidt, 'Social Functions of Emotions at Four Levels of Analysis', *Cognition and Emotion* 13 (1999): 505–21.

Keltner, Dacher, and Jonathan Haidt, 'Approaching Awe, a Moral, Spiritual, and Aesthetic Emotion', *Cognition and Emotion* 17 (2003): 297–314.

Kemal, Salim, and Ivan Gaskell, 'Nature, Fine Arts, and Aesthetics', in *Landscape, Natural Beauty and the Arts*, ed. Salim Kemal and Ivan Gaskell (Cambridge: Cambridge University Press, 1993), 1–42.

Ledley, Fred D., 'Visions of a Source of Wonder', in *Envisioning Nature, Science, and Religion*, ed. James D. Proctor (West Conshohocken: Templeton Press, 2009), 245–70.

Losch, Andreas, 'Our World is more than Physics', *Theology and Science* 3, no. 3 (2005): 275–90.

Matravers, Derek, 'Wonder and Cognition', in *Practices of Wonder: Cross-disciplinary Perspectives*, ed. Sophia Vasalou (Eugene: Pickwick Publications, 2012), 166–78.

McAllister, James W., *Beauty & Revolution in Science* (Ithaca and London: Cornell University Press, 1996).

McGrath, Allister E., *The Open Secret: A New Vision for Natural Theology* (Malden, Oxford and Carlton: Blackwell Publishing, 2008).

Moran, Joe, *Interdisciplinarity* (Abingdon: Routledge, 2002).

Mothersill, Mary, 'Beauty and the Critic's Judgment: Remapping Aesthetics', in *The Blackwell Guide to Aesthetics*, ed. Peter Kivy (Malden: Blackwell, 2004), 152–66.

Parsons, Glenn, *Aesthetics and Nature* (London and New York: Continuum International Publishing Group, 2008).

Polkinghorne, John, *Theology and Science* (London: SPCK, 1998).

Proctor, James D. (ed.), *Envisioning Nature, Science, and Religion* (West Conshohocken: Templeton Press, 2009).

Rolston III, Holmes, 'Does Aesthetic Appreciation of Landscapes Need to be Science-Based?' *British Journal of Aesthetics* 35, no. 4 (1995): 374–86.

Rolston III, Holmes, 'From Beauty to Duty: Aesthetics of Nature and Environmental Ethics', in *Environments and the Arts: Perspectives on Environmental Ethics*, ed. Arnold Berleant (Aldershot: Ashgate, 2002), 127–41.

Rolston III, Holmes, *Science & Religion: A Critical Survey* (West Conshohocken: Templeton Press, 2006).

Russell, Robert John, *Cosmology: From Alpha to Omega: The Creative Mutual Interaction of Theology and Science* (Minneapolis: Fortress Press, 2008).

Scarry, Elaine, *On Beauty and Being Just* (Princeton: Princeton University Press, 1999).

Seel, Martin, 'Nature: Aesthetics of Nature and Ethics', in *Encyclopedia of Aesthetics*, vol. 3, ed. Michael Kelly (Oxford: Oxford University Press, 1998), 341–3.

Shusterman, Richard, and Adele Tomlin eds., *Aesthetic Experience* (New York: Routledge, 2008).

Sobosan, Jeffrey G., *Romancing the Universe: Theology, Science, and Cosmology* (Grand Rapids: W. B. Eerdmans Publishing Co, 1999).

Stecker, Robert, *Aesthetics and the Philosophy of Art: An Introduction* (London, Boulder, New York, Toronto and Oxford: Rowman and Littlefield Publisher, Inc., 2005).

Stenmark, Mikael, *How to Relate Science and Religion: A Multidimensional Model* (Grand Rapids: W. B. Eerdmans Publishing Company, 2004).

Sæther, Knut-Willy, *Traces of God: Exploring John Polkinghorne on Theology and Science* (Trondheim: Tapir Academic Press, 2011).

Vasalou, Sophia (ed.), *Practices of Wonder: Cross-disciplinary Perspectives* (Eugene: Pickwick Publications, 2012).

Wilson, Ross, 'Beauty and Sublimity', in *The Oxford Handbook of Theology & Modern Thought* (Oxford: Oxford University Press, 2015), 419–34.
Zuckert, Rachel, 'The Associative Sublime: Gerard, Kames, Alison, and Stewart', in *The Sublime: From Antiquity to the Present*, ed. Timothy M. Costelloe (Cambridge: Cambridge University Press, 2012), 64–76.

Chapter 8

IMAGINATION AS CO-CREATION: SCIENCE AND THEOLOGY THROUGH THE LENS OF SCIENCE-FICTION LITERATURE

Zoë Lehmann Imfeld
University of Bern and University of Zurich

In his 1930 novel *Last and First Men*, British writer and philosopher Olaf Stapledon writes: 'Today we should welcome, and even study, every serious attempt to envisage the future of our race. … To romance the far future, then, is to attempt to see the human race in its cosmic setting, and to mould our hearts to entertain new values' (1930, 11). In science fiction, we create thought experiments in which we play out the consequences of the decisions we make today. As we increasingly aggressively pursue a group of sciences which brings not only our understanding but our physical presence further outside the bounds of Earth, this perspective is crucial if we are to continue to spread the footprint of humankind beyond our own planet. Science fiction is a medium which allows us to explore the ethical and metaphysical implications of this increasing footprint. These are narratives in which human beings participate in the continuing formation of the universe. Sometimes these are stories that tell of a teleological evolution, in which humankind grows into its place in the universe. Other texts serve as warnings, and depict human beings who imagine themselves as gods, and become destructive. Of course, science fiction itself often falls prey to the hubris and arrogance of imagining humankind as conquerors of space, but even such 'uncritical' approaches can prove useful texts for exploration, as will be seen in the texts under consideration here.

This chapter will focus on terraforming, a term coined by Jack Williamson in his short story 'Collision Orbit' from 1941. Terraforming, the altering of a hostile terrain or biosphere to accommodate the conditions for life, in fact has a long tradition in science fiction. In *The War of the Worlds* (1898), H. G. Wells's Martians begin to alter Earth for their own purposes during their brief occupation, and humans terraform Venus in Stapledon's *Last and First Men*. In the scientific climate of recent decades, terraforming would seem to offer a 'second chance' for a human population whose destruction of our own planet seems increasingly irrevocable. This is the case in Frederick Turner's *Genesis*, discussed later in the chapter, in which humankind terraforms Mars to escape a ravaged Earth. To Turner, Mars is an untapped resource waiting for humankind to come and unlock its potential.

The terraforming of Mars is not only about leading humankind into its next stage of evolution, but also about turning Mars from a dormant to a fertile planet (by releasing deep reserves of oxygen from beneath its surface). By terraforming Mars, human beings are literally breathing life into their creation. Indeed, terraforming is not entirely restricted to the imagination of science-fiction writers. In 1976 NASA published a technical report entitled *On the Habitability of Mars: An approach to planetary ecosynthesis*, which described the artificial creation of an oxygen-based ozone as feasible. In terms of the 'real life' viability of such a project, the NASA-based Mars Ecopoiesis Test Bed project recently simulated the conditions of Mars to explore the possibility of microorganisms being used to draw nitrogen from the Martian regolith. The project aims to select a site from which Martian soil can be extracted and tested.[1] Even more ambitious are the activities of Elon Musk's SpaceX company towards human habitation on Mars (in Musk's vision by 2024, see Musk 2017; see also Persson, this volume). Of most interest here, however, is what the *imagined* potential of terraforming signifies. Narratives of extra-planetary terraforming, both fictional and theoretical, are an embodiment of the imagined potential for human scientific evolution. Terraforming raises questions of humankind's ethical relationship with its environment, both on Earth and into the solar system. (Indeed, these are many of the questions which are considered in other chapters of this volume.) For author Kim Stanley Robinson, for instance, terraforming suggests redemption, a second chance for humankind as planetary stewards (Robinson *Remarks*, 9). To Robinson, 'the idea of science as a utopian coming-into-being has seemed both true and useful, suggestive of further stories and action in the world' (4). Such a utopian outlook, however, also demands reflection on our current planetary stewardship. Planetary habitation can be symbiotic or exploitative. The novel *Red Mars*, discussed later here, paints a clear parallel between the use of Mars as a mining resource, and the potential use of Antarctica for the same reason. (Within the novel, Antarctica has become a mining resource.) Although the novel was published shortly after the Madrid Protocol to prohibit Antarctic mining was signed in 1991, the novel reflects anxieties that the weight of the prohibition ensured more by economic impracticality than by ethics.[2] For Robinson, our stewardship of the Earth is indicative of our stewardship of the solar system and beyond.

Depictions of human beings as causing fundamental changes to the planets in the universe demand interrogation of our role in the creative process of the universe. In each case, human journeys into space, and the act of creating habitations for themselves, are intrinsically bound with *becoming*. The development of human activity in the cosmos that science fiction imagines is theologically complex. Some texts describe extraterrestrial expansion as teleological (à la Whitehead), a

1. See https://www.nasa.gov/content/mars-ecopoiesis-test-bed.

2. Robinson would go on to explore these themes more fully in his 1997 novel *Antarctica*, a near-future novel in which the area has come under threat from the exploitation of natural resources.

fulfilment of human potency. In some narratives we see a creative move towards fuller participation in the universe as a communal organism (à la Teilhard). In other texts, however, this creative impulse is destructive – humankind comes to think of itself as godlike, in wanton disregard for natural law (à la Augustine and Thomas).

The aim of this chapter is not to promote a particular theological position on space sciences. I will suggest, however, that aspects of process philosophy and process theology provide a rich vocabulary through which to interpret the inherent dynamism of an imagined future. The structures of Thomist natural law, however, are shown to provide checks and balances to the dynamic optimism of process thought. For process philosophers, becoming is central to being, and they insist that reality itself is continuously coming into being. Early process philosophers, such as John Dewey, George Herbert Mead and Alfred North Whitehead, sought to account for the novelty in nature that was described by Darwinian evolution. The question arises then, of whether this novelty is directed or self-sustaining, a question which would lead to a split between naturalist (often described as secular) process thought and teleological (and generally theist) (see Reschner 2006). For Teilhard de Chardin, for instance, evolution is a demonstration of God as '*Prime Mover*' (Teilhard 1969, 28). As 'Prime Mover', 'God *does not make:* He *makes things make themselves*' (28, original emphasis). In this way, process theists can think of God and beings as co-creators (Hartshorne 1967, 113; Hartshorne and Reece 1953, 140). This doctrine of creativity differs from classical theism which states that God alone is creative. Aquinas, for instance, understands creation as to bring something from nothing, ex nihilo, and thus only possible of God (*Summa* I, q. 45, a.5). The role of humankind then, is to come fully into being by *participating* in God's creation. Science fiction presents us with narratives of human development (technological and biological) that can be described in terms of Aristotle's metaphysics of potency and act (*Metaphysics*). In these texts we therefore see also reflections of the theological anthropologies of SS Augustine and Aquinas. As theological beings, created in the image of God, human beings contain the potential to *become* the fullest theological version of themselves, and moreover, to participate in the universe as creation.

The tension created by these perspectives is repeatedly played out in science-fiction narratives of space exploration and habitation. By giving humankind a creative relationship to the universe, the science-fiction texts under consideration here explore questions concerning the opportunities for human actualization that the universe beyond earth affords. Also what, if any, role humankind has in releasing the potency of the universe. Perhaps these perspectives can be reconciled. Is the expansion into space an act of co-creation – human beings participating in bringing the universe more fully into being? Or, by behaving as god like creators, are human beings ultimately destructive – spreading parasitically into space having exhausted the resources of earth?

This chapter will follow three science-fiction texts in order to explore how depictions of terraforming reflect the development of space sciences as ethics, through the late twentieth and early twenty-first century. The texts are not explicitly Christian or theological fictions, but each imagines a future for humankind that challenges notions of human being and becoming in the universe. As we shall see, science fiction allows us to 'entertain new values', and to explore a role for human beings as participants in a process of co-creation, but it also recognizes potential limitations. Science fiction asks: How far can humankind evolve and still remain human? Which version of human evolution can be a creative force in the universe and still remain the *imago Dei*?

Frederick Turner's Genesis and the Gaia Hypothesis

As is central to the themes of this volume, the question of humankind's relationship with our environment beyond earth is primarily an ethical one. So too are the narratives of becoming depicted in science fiction. The first text under consideration, the epic poem *Genesis* from 1987, demonstrates the inextricable relationship between ethics and becoming, by presenting a naïve and essentially 'un'-ethical ideal of human expansion into space. Styling itself in the tradition of epic poems as creation stories, this is a poem in which an uncritical approach to humankind as creator is ultimately unconvincing both literarily and philosophically. This three-hundred-page poem opens on a near-future earth so abused by humankind that it has become a desolate landscape. Both nature and humankind are sick and dying: 'Shadows of men. / They are the makers of / This desolation, these cocytean swales./ What must they be to be guilty of such / An ugliness?' (I.i, 87–90). The world is governed by an ecotheist regime which sees humankind as evil in its exploitation, and Earth's environment as its victim. Untouched and unspoiled nature is deified, and society is regulated by maintaining undisturbed nature at all costs. The tenets of this regime are founded in a return to (Thomist) medieval understandings of natural law, and the claim 'Such human arrogance, Lords of Creation / As we thought ourselves to be, was the cause / Of ecocidal holocausts before' (II.iii, 21–4).

Aquinas deals with eternal, natural and human law in his *Summa Theologica*, and these laws are concerned with creation and God's conception of it. Aquinas describes 'eternal law' as the plan of God in creation, with God as the exemplary 'template'. The more a thing participates in this template, the more complete it is. Natural law, then, is a way of being which participates in this eternal law. To the ecotheists of *Genesis*, humankind's destruction of their environment has contravened the 'natural law' of being for humans, as it is destructive to God's eternal plan. The regime has a group of rebels, however, led by the poem's hero named Chance, who leads a group in the prohibited task of terraforming Mars. The poem thus tells the tale of the battle between a stagnant theist 'old world order' on Earth, and the human inheritors of the universe, destined to fulfil 'the meaning of the race, / Which is to make – to master, mother, make' (II.ii, 344–5).

Genesis, as Turner describes in his preface, is heavily influenced by the Gaia hypothesis, made popular in the 1970s by James Lovelock. According to this hypothesis, the Earth is a self-regulating system, in which each of the Earth's atmospheric spheres are part of a linked and evolving system that maintains homeostasis (see Lovelock 1995).[3] The suggestion in *Genesis* is that the terraforming of Mars is something akin to a 'second round' of the evolution of Earth's homeostasis. As is clear from the preface, the potential for an accelerated planetary evolution based on the Gaia hypothesis seemed not only plausible to Turner, but technologically just around the corner. He mentions evidence of water having once existed on Mars, and the potential for computer sequencing to 'construct' biology (using the then potential DNA sequencing as an example) ('Explanatory and Reference Material' 4, 5). In Turner's vision, terraforming of Mars brings humankind into biotic symbiosis with the universe. Turner's narrative of Mars's terraformation is heavily teleological, both for humankind and for Mars, and it is possible to detect notions of Teilhard's *Omega* point. In the *Omega* point, the Jesuit-palaeontologist saw a reconciliation between biology and Christianity by describing evolution in terms of orthogenesis. For Teilhard, all participants in the universe are necessarily convergent in nature, moving towards (or rather evolving towards) a cosmic consciousness (the *Omega*, or God). Teilhard writes: 'In the perspective of a noogenesis, time and space become truly humanised – or rather superhumanised. Far from being mutually exclusive, the Universal and Personal … grow in the same direction and culminate simultaneously with each other' (285). In Turner's *Genesis*, this convergence of human consciousness into cosmic consciousness goes hand in hand with the process of terraforming Mars.

Crucially, this cosmic teleology is intimately bound with creativity, namely the *creation* of Mars. The 'making of Mars into a paradise' enables for its makers 'transubstantiation into men' (I.i, 397–9). Turner makes parallels between the act of writing and terraforming. Mars is a 'land void of inscription, storyless', with soil that 'must be fattened with gifts of tongues and visionary blood' (IV.iv, 221, 233). Mars's teleology is therefore as dependent on humans as human teleology is on Mars. The purpose of Mars is to be brought into being *by* humans. It is this dependence which ultimately makes Turner's convergent teleology problematic.

As the poem progresses, Turner's narrative becomes less one of human co-creation, and increasingly of Creation. Turner likens the terraforming of Mars to a sort of 'Genesis 2', and his eponymous poem to a 'revelation special and proper to Mars, as the sayings of Isaiah, Jesus, Buddha, Mohammed, Lao Tse and the others were revelations proper to the Earth' (V, introductory lines). Ultimately, there is no theological communion, as Turner suggests, but rather

3. Lovelock's 1985 novel *The Greening of Mars*, co-written with Michael Allaby, uses the terraformation of Mars to promote the Gaia hypothesis. Robert Markley describes the novel thus: 'As a thought experiment, the novel constitutes the ideational ground – the values, assumptions, and theories – upon which the emerging discipline of planetary engineering rests. More generally, it suggests the extent to which the science of terraformation relies on mythic archetypes of resurrection for its rationale' (1997, 778).

dominion, humankind becoming the gods of the solar system: 'He will have no other god before Him/ And every human maker is a god' (III.iv, 280–1). In many ways, the fears of the ecotheist regime come true. Human beings have evolved not into co-creators but have *replaced* a creator God-figure. 'That god', we are told, 'remains a ghost beyond the living walls of nature' (V.i, 250–1). There is, however, no ethical or theological consequence for this radical reconfiguration of natural law, but rather the actualization of both humankind and Mars. Turner leaves an exploration of the ethical implications for a human 'footprint' on Mars to emerge more fully in the science-fiction novels of the 1990s.

Kim Stanley Robinson's Mars Trilogy

Questions of symbiosis also infuse Kim Stanley Robinson's trilogy of terraforming novels, published between 1993 and 1996, although these are arguably more ethically sophisticated. The first novel of the trilogy, *Red Mars*, begins with an international expedition on its way to Mars to start the process of terraforming. As well as various national and cultural groups, the crew members represent various goals for the expedition, some seeing Mars as a potential mining tap for natural resources, some as a new home for humankind, and others as an object for scientific study. Two dominant themes preoccupy the members of its crew as the novel progresses: First, the ethical responsibility of humans towards Mars, and second, the consequences of living on Mars for what it means to be human. Both of these themes, as we shall see, become emblematic of what Solomon Katz and theologian Philip Hefner describe as 'bio-cultural evolution' (Katz 1980; Hefner 1993, 29sq.). As the trilogy develops through terraforming, the establishing of communities, civil war and the emergence of its own set of myth's and rituals, the novels are as much a narrative of social/ethical evolution as they are a depiction of the transformation of Mars. An evolutionary model such as Hefner's requires the concept of both genetics and extrasomatic information as having evolutionary functions. To Hefner, bio-cultural evolution is central to the fulfilment of our theological role as co-creators in the universe. In Hefner's theological model: 'The freedom that marks the created co-creator and its culture is an instrumentality of God for enabling the creation (consisting of the evolutionary past of genetic and cultural inheritance as well as the contemporary ecosystem) to participate in the intentional fulfilment of God's purposes' (1993, 32). Culture and nature, then, provide a cyclical model for human evolution. Both genetics and culture provide the direction for humankind's continuing development.

Participating in co-creation, then, is integral to humankind's process of becoming. The first novel in the trilogy, *Red Mars*, examines this process in the context of terraforming, the characters recognizing within the opening chapters that by bringing change to Mars, they are bringing change upon themselves. However, the text complicates the notion that habitation on Mars is a neutral and inevitable step in human evolution. The crew recognizes that such an evolution requires human participation in the process. As they travel towards Mars, engineer and political anarchist Arkady stirs up dissent among the crew with the following speech:

> We can do this because we have technology to manipulate matter right down to the molecular level. This is an extraordinary ability, think of it! And yet some of us here can accept transforming the entire physical reality of this planet, without doing a single thing to change ourselves, or the way we live. … It's absurd, it's crazy, it's – it's – ' he seized his head in his hands, tugged at his hair, roared 'It's *unscientific*! And so I say that among all the many things we transform on Mars, ourselves and our social reality should be among them. We must terraform not only Mars, but ourselves. (112)

Unlike the members of *Genesis's* Gaian paradise, however, Robinson's characters are confronted with the ethical consequences of initiating such a change. As the project progresses, the crew find themselves split into two groups. The first, who become the 'Red Movement', are led by the geologist Ann Claybourne in the belief that the *only* ethical role for humans on Mars is in 'hands-off' scientific study. To change the planet is to destroy it. Terraforming is both unscientific and immoral (204). An opposition group develops in what becomes the 'Green Movement', led by Sax Russel. Sax's outlook is deeply anthropocentric, reminiscent of the ideology that fuels *Genesis*:

> 'The beauty of Mars exists in the human mind', [Sax] said in that dry factual tone, and everyone stared at him amazed. 'Without the human presence it is just a collection of atoms, no different from any other random speck of matter in the universe. It's we who understand it, and we who give it meaning. (212)

Such a stance, indeed, seems to fit neatly with the political background of the expedition, in which hyper-capitalist interests view Mars purely as a resource to be exploited. Inevitably, the rapid change imposed on the Martian environment (heating of the atmosphere and melting of permafrost) leads to civil war and violent climate change, and *Red Mars* finishes among avalanches and floods. This instability continues through the second novel (*Green Mars)* and is only partially resolved in the final novel, *Blue Mars*.

These changes and eventual equilibrium occur through economic and political actions, but the changes to humankind are far more profound. In the context of human teleology/co-creation, I would argue that the two ethical positions depicted in the first novel are *both* fundamentally anthropocentric.[4] While Sax and his followers may claim dominion over Mars, Ann sees humankind only as observers. 'Life', for Ann, is something other than humans. To Ann, any contact by humans with Martian life would, by changing it, essentially destroy

4. For a practical discussion of the moral issues of anthropocentrism in the solar system, see Persson, this volume.

what it is. Ann's philosophy therefore affords no process of symbiosis between humankind and the universe, humankind inhabiting a unique, privileged place in the universe. To Ann, humans have a *separate* status, in a way that subverts but ultimately reinforces anthropocentrism. It is this anthropocentrism which Robinson truly challenges in the *Mars Trilogy,* and overcoming it becomes central to human evolution. As in *Genesis,* notions of a human convergence with the universe emerge in the *Mars Trilogy*. However, Robinson introduces ethical checks and balances that go unexplored in *Genesis*. At either philosophical extreme (That of the 'Red' or the 'Green' movement), humankind's failure to recognize themselves *as part of* their environment leads to disaster, both social and environmental.

Environmental equilibrium is eventually achieved through a model of 'eco-economics', an ecologically based value system which is developed by two biologists in *Red Mars* (351) and implemented in the final novel, *Blue Mars*. However, another change occurs between these two novels, represented by the character Hiroko Ai, *Green Mars*'s teacher of the first-generation Martian children. By *Green Mars*, society has been fractured by civil war, and Hiroko leads a small, utopian community in an isolated ice-dome. Hiroko leads this group in an ecotheist philosophy called viriditas, which seeks to unify microcosm with macrocosm:

> Look at the pattern this seashell makes. The dappled whorl, curving inward to infinity. That's the shape of the universe itself. There's a constant pressure, pushing toward pattern. A tendency in matter to evolve into ever more complex forms. It's a kind of pattern gravity, a holy greening power we call viriditas, and it is the driving force in the cosmos. Life, you see. … And because we are alive, the universe must be said to be alive. We are its consciousness as well as our own. We rise out of the cosmos and we see its mesh of patterns, and it strikes us as beautiful. And that feeling is the most important thing in the universe – its culmination, like the color of the flower at first bloom on a wet morning. It's a holy feeling, and our task in this world is to do everything we can to foster it. (*Green Mars* 19)

This passage is important, as it demonstrates the way in which Hiroko draws together the teleological claims of each of the approaches towards terraforming, and places them in the context of process theology. The passage acknowledges Sax's privileging of human consciousness ('we are its consciousness'), and also Ann's insistence that we are only a small part of the cosmos ('we rise out of the cosmos'), all within the context of a biological definition of evolution ('a tendency in matter to evolve into more complex forms'). Crucially, however, Hiroko places the will of tendency not on humankind, but on the universe. Human beings are not simply working on the universe, but are being worked upon.

Human beings are also physically changed by Martian habitation, and they participate in the acceleration of this change too. For instance, the members of the Martian community make use of genetic engineering which dramatically increases longevity. It is, however, cultural and philosophical changes which allow for the fruition of these changes. Once again, we see Hefner's model of bio-cultural evolution depicted. Hefner writes: 'It is science that sets forth the fundamental descriptions of our human teleonomy, but it is myth and ritual that make the basic proposals concerning the direction, meaning, and purpose of the structures and processes whose fulfilment shapes the form of human being' (1993, 21). In the *Mars Trilogy,* this fecund interaction between science and myth is embodied in the characters, namely Hiroko. Beginning as a geneticist, Hiroko goes on to develop and promote a symbiotic agronomy for humans on Mars, which becomes the mystic philosophy described above. Indeed, Hiroko herself attains a mythical status among the other characters, and her death is followed by reported sightings and rumours of her survival. Hiroko's influence on the ecological and economic ethics of the trilogy is played out through her influence on Sax and Ann, both directly, in discussion and argument, and indirectly, such as Hiroko's mythological presence in the final novel. (For instance Sax believes he is rescued by Hiroko from a snow storm.) The crew's psychologist Michel Duval describes it thus: 'In archetypal terminologies we might call green [Sax's *Weltanschauung*] and white [Hiroko's *Weltanschauung*] the Mystic and the Scientist. Both extremely powerful figures, as you see. But what we need, if you ask me, is a combination of the two, which we call the *Alchemist*' (*Green Mars* 15, original emphasis). Again, Robinson's nomenclature is important. The 'green' will to create, given impetus by the 'white' teleonomy, becomes the alchemist – creator of something new through material change.

By the final novel, *Blue Mars,* ideological oppositions break down, and a degree of environmental equilibrium is achieved. The inhabitants of Mars become 'Martian' humans. Robert Markley makes the claim that the equilibrium achieved by *Blue Mars* comes about because eco-economics triumphs over spiritual and ideological approaches, what Markley calls an 'authorial endorsement of eco-economics' (1997, 792). Certainly, *Blue Mars* sees the breakdown of ideological oppositions and the implementation of an eco-economic society. However, the *change* that takes place, particularly within the emblematic characters of Sax and Ann, is that the understanding of human teleology moves from being anthropocentrically driven to being cosmically driven. Ecological/human symbiosis is achieved not simply in the sense that ecological stability is restored, but because human beings have been *worked on* by the process of terraforming Mars. The fledgling Martian community both creates and is created by terraforming. For Kim Stanley Robinson, terraforming is redemptive – faced with the looming crisis of Earth ecology, terraforming offers the chance for humankind to respond with creativity (see *Future Primitive* 11). Both the redemption and the process of change is, however, an opportunity *afforded by* the universe. Creativity is given as a redemptive opportunity. We again see something Teilhardian, then, about Robinson's *Mars Trilogy* – the primacy of individual will and consciousness recedes to biosynthesis with humankind's environment. In this case, the 'individual' is anthropocentric humankind. Robinson's trilogy reinforces the notion that by changing the universe, humankind is changed. This does not address the question, however, of what it means for the 'being' of human being to be changed. Christian theology describes human being as the *Imago Dei*, the image of God. How far can human being change from its own image before this comes into question?

Adrian Tchaikovsky's Children of Time

The process depicted in the two previous texts is in many ways concerned with return, the restoration of equilibrium. Both Mars and humankind grow into the fulfilment of their teleologies. Tchaikovsky's 2015 novel *Children of Time* also includes terraforming as a pivotal step in human process and evolution, but on a far larger – interstellar – scale, and encompassing millennia. Humans terraform suitable planets and implant them with organisms ready for accelerated evolution. The result of this evolution is at once recognizable and entirely new, and Tchaikovsky's text problematizes the meaning of 'potency' at the heart of process theism, by challenging the idea of a movement towards a fixed, ideal 'form'. *Children of Time* is also further complicated by seeming to at once reinforce and shatter a Teilhardian notion of convergent evolution.

Children of Time contains two parallel narratives. The first begins with a far-future project to terraform exo-planets and plant on them the seeds for continuing evolution, the brainchild of Doctor Avrana Kern. The novel opens on the 'Brin 2' facility, orbiting a recently terraformed exo-planet. The facility carries primate specimens and a genetically engineered virus, designed to work as a catalyst for the accelerated evolution of the primates when both are deposited on the planet. The continuing survival of humanity is to be ensured by the introduction of this nano-virus which will direct the evolution of the planet's primate inhabitants: 'No matter how much change the monkeys underwent, that virus would adapt and adjust to whatever genome it was partnered with, analysing and modelling and improvising with whatever it inherited – until something had been engineered that could look its creators in the eye and understand' (8). The project appears to be the ultimate combination of teleonomy and creativity. Just as the terraformers of *Genesis* are hampered by an ecotheist regime, however, this project too falls victim to an opposition group, in this case in the form of a terrorist attack just as the project begins: 'You have to be stopped, Doctor Kern. … It ends here. *Non Ultra Natura!* No greater than nature' (11). When the facility is destroyed in this attack, instead of a carefully controlled experiment, the terraformed planet and its nano-virus is left to evolve alone, orbited only by Avrana Kern as its sentinel, in the deep hibernation of a survival pod.

The second narrative joins an interstellar ark-ship several thousand years later, carrying the remnants of humanity who have fled Earth after wars and destruction rendered it uninhabitable. The crew of this ship, the Gilgamesh, come across this apparently habitable planet and recognize it from legends of 'ancient' terraforming projects. These narratives collide as the humans of the Gilgamesh and the evolved inhabitants of the terraformed planet are confronted with each other. The additional challenge is that in the chaos of its inception, the nano-virus has taken root not in primates, but in arachnids. The inhabitants are not humanoid 'grandchildren' of Earth, but giant spiders.

In this way *Children of Time* poses radical challenges to the relationship of cosmological creativity and purpose, and to forms of being and their relationship to the image of their creator. How far can humankind participate in the cosmological process, and still claim it as its own process? Here Henri Bergson's response to process theology proves useful in its reconsideration of the nature of creative evolution. Freedom of direction is central to Bergson's model of creative evolution. For Bergson, cosmological teleology precludes creativity, since it entails that the 'whole is given'. This is a direct challenge to Augustinian interpretations of potency, or to the fulfilment of a natural law. The leader of the Gilgamesh ark-ship epitomizes Bergson's model when asked about his plan to land on the terraformed planet: '"Are you proposing this as the future of the human race?" Vitas asked flatly. "*The* future, no. *A* future, yes"' (116).

For Bergson's model this connection between creativity and newness also informs our relationship with form. While we look to form as the essence of a thing, to Bergson a thing's essence is in the changes that it undergoes:

> Now, life is an evolution. We concentrate on a period of this evolution in a stable view which we call a form, and, when the change has become considerable enough to overcome the fortunate inertia of our perception, we say that the body has changed its form. But in reality the body is changing form at every moment; or rather, there is no form, since form is immobile and the reality is movement. What is real is the continual *change of* form: *form is only a snapshot view of a transition*. (Bergson 1911, 302, original emphasis)

The narrative structure of *Children of Time* means that the human beings on the Gilgamesh are confronted with a different snapshot of human evolution than their own. The crew's ancestors had initiated a stream of evolution through terraforming, but the results are unrecognizable. The crew of the Gilgamesh is confronted not with 'Planet of the Apes'-style primates, but with monstrous sentient spiders:

> It happened very swiftly. One moment they were staring out into the clear space ahead that the drone was being inexplicably denied, and then a vast hand-like shadow eclipsed their view. They had a moment's glimpse of many bristling legs spread wide, two fangs like curved hooks striking savagely towards the camera with ferocious speed and savagery. …
>
> 'That was no fucking monkey', Karst spat. (Tchaikovsky 2016, 118)

Yet these arachnids, thanks to the inherited nano-virus, are also in many ways profoundly human – in their social hierarchies, in their scientific practices and in their myths and rituals. Tchaikovsky challenges the premise that the crew of the Gilgamesh are fundamentally 'more' human, and the spiders an aberrant evolutionary divergence.

Tchaikovsky is able to challenge this anthropocentric perspective through the character of Avrana Kern. As an emergency response to the initial terrorist attack on the terraforming project, Avrana placed herself in a hibernation capsule orbiting the terraformed planet, and connected herself bio-electronically to the survival pod's AI system. Millennia later, Avrana remains isolated from both the inhabitants of the terraformed planet, and the humans of the Gilgamesh, being the last of her own generation of humans. She is physically and mentally 'outside' either one of the text's narratives. Her position exceeds the temporal 'snapshot' of either narrative. From this perspective, Avrana responds very differently to the two evolutionary streams she encounters. Seeing herself as the creator of the terraformed planet, she views its inhabitants as the inheritors of humankind: 'She sees them now, yes. She sees them for what they are. They are Earth. Their form does not matter. They are her children' (456). Meanwhile, despite the Gilgamesh's pleas to land on the terraformed planet, Avrana does not recognize them as human beings: 'I do not recognise you. You are not human. You are not from Earth. You have no business here. Eliza [the AI system] shows me all that she sees of you and there is nothing of Earth in you' (92).

Bergson, arguably, would claim that Avrana has (partially at least) recognized the 'real duration' of co-creation, signifying 'both undivided continuity and creation' (Bergson 1911, xiv, n1). It is this 'duration' in which human beings are usually unable to participate: 'Instead of attaching ourselves to the inner becoming of things, we place ourselves outside them in order to recompose their becoming artificially' (306). Indeed, it is this sense of continuity and creation which carries the two narratives of *Children of Time* through to the novel's conclusion. The theme of inheritance and generational memory is strong in both narrative streams, but with very different directions. On the Gilgamesh ship a classicist, Holsten, repeatedly wakes from a form of cryogenic hibernation to find that generations have passed on the ship. With each generation he finds the communal memory (for instance of Earth) deteriorating, and is burdened with the realization that he alone as a historian continues to carry it: 'He shivered – not from the usual human sense of mortality, but from a feeling of vast, invisible things falling away into oblivion, irretrievable and irreplaceable' (500–1). The contrast of this is stark against the arachnid narrative. We are introduced to the protagonist of this narrative, Portia, through the genetically inherited behaviours typical of a spider species. The implanted nano-virus, however, has also begun to transcribe human behaviour onto the genomes of each generation, passing on not only knowledge ('ragged-edged with loss of context'), but a strengthening of communal cooperation and identity: 'The economic, force-evolved brains of Portia's kind share more structural logic with each other than chance-derived human minds do' (70). By accelerating evolution, this human interference has also accelerated the subsuming of individual to communal consciousness. This is subtly reinforced as it slowly becomes clear that the narrative is not following 'Portia', a single protagonist, but 'Portia' as a generational series of inherited memes.

It is in this way that *Children of Time* fulfils the interdependency of continuation and creativity. Just as in *Genesis* and the *Mars Trilogy*, human participation in co-creation goes awry at the point that they stop behaving as co-creators and begin to think of themselves as Creator. As with our previous texts, this is expressed in the form of a 'god-complex' (both Avrana and the leader of the Gilgamesh come to think of themselves as the god of their domains). However, in *Children of Time* it is also expressed in a more complex form: The more that creative evolution is defined by 'purpose' or 'direction', the more stagnant it becomes. For the crew of the Gilgamesh, refugees from an uninhabitable Earth, the purpose of the mission is to find a new planet on which to settle and ensure the continuation of humankind in the universe. The more desperate the situation seems to become, the more Guyen, the ship's leader, comes to see himself as some kind of saviour. He begins to encourage a legend of himself as a messiah-figure, returning intermittently from hibernation to a new generation of crew. Yet within the confines of the ship, and the confines of Guyen's plans, there is no change, no creativity:

> 'I will provide for them', Guyen promised. 'I will show them the way.'
>
> 'It'll be the moon colony all over again', Holsten snapped. 'They'll die. They'll eat all the food. They'll just … *live* everywhere until things break down. This isn't a cruise ship. The *Gil* isn't supposed to be *lived* in. They're cargo. We're all cargo.' (401)

Again, this theme is reinforced by contrasting it with the parallel narrative. At its inception, the terraformed planet and accelerated evolution is designed to be tightly controlled. Even as it is conceived, its purpose is profoundly anthropocentric. The nano-virus, for instance, 'was designed that way in order to creatively accomplish its hardwired aim: to bring the host to a detected level of sophistication set by its creators and, once its victory conditions are met, to cease further assistance. Its creators included such safeguards so as to prevent their protégés continuing to develop into superhuman monkey-gods' (69). Humankind as creator cannot envisage a creation beyond its own image. Avrana, as the last human orbiting her terraformed creation, at first sees herself very much in these terms. As the inhabitants of the planet reach a certain level of technological sophistication, Avrana is able to send them messages via the AI system. 'Tell them this': Avrana instructs the AI, 'I am your creator. I am your god' (310). The relationship continues on these terms for much of the narrative, as the arachnids mythologize and deify Avrana, calling her 'The Messenger', and establishing a strict religious orthodoxy based on attempts to interpret her (difficult to translate) messages. Crucially, however, the community do not see The Messenger as a saviour, but only as evidence that there is more, 'a logic to the world, extending beyond the mere chaos of the physical' (158). Two things happen that allow creativity to flourish among the arachnid community: First, a gentle de-consecration of The Messenger and her role. Secondly, Avrana comes to terms with the realization that the inhabitants of her created planet are not the 'sentient monkeys' that she had envisioned, and that their purpose need not be her own. By the time their technology allows two-way communication with Avrana, her new message to them is this: 'You are mine, but you also belong to the universe, and your purpose is whatever you choose' (481).

In many ways then, *Children of Time* is profoundly anti-eschatological, demanding that cosmological creativity requires duration but also novelty. This would seem to be a direct challenge to the concept that human beings can participate in co-creation of the universe by 'unlocking' their own and the universe's potency, and bringing it into actuality. However, a sense of convergence and interconnectivity does remain in the conclusion of this text. When the Gilgamesh attacks the inhabitants of the terraformed planet, in a final attempt to find a place to settle, the arachnid population responds with chemical warfare – not of poison, but by unleashing the nano-virus back onto the humans. The evolution of the spider population has been the triumph of self-recognition (they have identified and replicated their own genetic inheritance, calling it 'the Understanding') and the triumph of interconnectivity, the spiders responding with a form of recognition when faced with the 'alien' humans. Once the creation of humankind, the arachnid population now in return provide the catalyst for the continuing process of human evolution:

> The virus in them is telling them all, *This is us; they are like us*. It tells the spiders the same, that crippled fragment of virus calling out to its more complete cousins: We are like you.
>
> And [Avrana] guesses, then, that the spiders' meddling might go further than they had thought. If there had been some tiny bead present in the brain of all humans, that had told each other, *They are like you;* that had drawn some thin silk thread of empathy, person to person, in a planet-wide net – what might then have happened? (592)

It is clear that the anthropocentric efforts of Avrana's terraforming generation and the stagnant refugee status of the Gilgamesh's crew were both doomed. The survival of the 'being' of humans is possible only in the novelty of the post-human arachnids.

Conclusion: Escaping the Snapshot View through Science Fiction

Physicist John Jungermann distinguishes between two types of chaos: deterministic and random (134). In deterministic chaos, the solutions are calculable in principle, but these calculations are impossible due to the almost infinite number of solutions. Random chaos, on the other hand, has no organization. This, Jungermann explains, 'is the world predicted by the second law of thermodynamics of maximum entropy and by "heat death"' (2000, 134). The apparent chaos of cosmological process cannot, then, be random, and it is this that affords a process theology of the universe. Some go on to claim intelligent design, and that co-creation involves discerning that design, and discerning humankind's place in it. The difficulty, however, is in seeing this 'process' from outside our own snapshot of it. As the narratives of science fiction repeatedly warn, believing that we can see and manipulate this 'design' is to see humankind not as co-creators but as creators.

In our three texts, those voices of caution against this human dominance over the universe come in the form of natural law. From a Thomist perspective, natural law concerns not our knowledge of the universe, but the nature of *how* we know it. Our mistake in claiming to see a design for the universe, or worse, to have dominance over that design, is to misunderstand the nature of human being. Aquinas goes so far as to call this demonic, as 'a false practical estimate of what is the true good' (*Summa* I, q.54, a.5). Our texts, however, while each responding in different ways to this caution, all reveal a more complex aspect of human creative fallibility. In each case, the failure of humankind's attempts at co-creation demonstrates a failure to be worked upon, to become changed. The 'god complex' exhibited in each of the texts is in fact an expression of the idea that humans can change their environment without being themselves subject to change by it.

For Bergson, we cannot participate fully in the 'process' of the universe, as we are trapped within a snapshot of it. We can apprehend the changes that occur because of that process, but not the process itself. This, says Bergson, is the 'natural law' of human being in the universe: 'Our intellect, in the narrow sense of the word, is intended to secure the perfect fitting of our body to its environment, to represent the relations of external things among themselves – in short, to think matter' (ix). Science fiction moves our perspective out of this snapshot, and allows us to view the process *as* process. As Bergson writes: 'The trick of our perception, like that of our intellect, like that of our language, consists in extracting from these profoundly different becomings the single representation of becoming *in general*, undefined becoming, a mere abstraction which by itself says nothing and of which, indeed, it is very rarely that we think' (1911, 304). Science fiction provides what Robinson calls 'historical simulations, which start at the present and then state *if we do this we will reach here*, or *if we do that we will reach there*' (*Future* 9, emphases in the original). This is a fundamentally utopian perspective, says Robinson, as it assumes that our actions now will have predictable consequences. However, as the texts presented here have demonstrated, it serves just as well to critically interrogate the narratives that we create for our own role in the universe. Naïve utopianism, both in science fiction and the natural sciences, is only a step from megalomania. Only through critical engagement will it become clear whether we do, in fact, have a role to play in the continuing process of the cosmos.

References

Aquinas, Thomas, *The Summa Theologica of St. Thomas Aquinas*, trans. by Fathers of the English Dominican Province (London: Burns Oates and Washbourne, 1915).

Aristotle, *Metaphysics*, trans. George Whalley (Montreal: McGill-Queens Press, 1997).

Averner, Maurice M., and Robert David MacElroy, *On the Habitability of Mars: An Approach to Planetary Ecosynthesis* (Washington DC: National Aeronautics and Space Administration, 1976).
Bergson, Henri, *Creative Evolution*, trans. by Arthur Mitchell (New York: Henry Holt & Co., 1911).
Hartshorne, Charles, *A Natural Theology for Our Time* (La Salle: Open Court, 1967).
Hartshorne, Charles, and William L. Reece, *Philosophers Speak of God* (Chicago: University of Chicago Press, 1953).
Hefner, Philip, *The Human Factor* (Minneapolis: Fortress Press, 1993).
Jungermann, John A., *World in Process: Creativity and Interconnection in the New Physics* (Albany: State University of New York Press, 2000).
Katz, Solomon, 'Biocultural Evolution and the Is/Ought Relationship', *Zygon: Journal of Religion and Science* 15 (1980): 155–68.
Lovelock, James, *Ages of Gaia* (Oxford: Oxford University Press, 1995).
Lovelock, James, and Michael Allaby, *The Greening of Mars* (Boston: Warner Books, 1985).
Markley, Robert, 'Falling into Theory: Simulation, Terraformation, and Eco-Economics in Kim Stanley Robinson's Martian Trilogy', *Modern Fiction Studies* 43, no. 3 (1997): 773–99.
Mead, George Herbert, *Movements of Thought in the Nineteenth Century*, ed. M. H. Moore (Chicago: University of Chicago Press, 1936).
Musk, Elon, 'Making Humans a Multi-Planetary Species', *New Space* 5, no. 2 (2017): 46–61.
Reschner, Nicholas, *Process Philosophical Deliberations* (Frankfurt: Ontos Verlag, 2006).
Robinson, Kim Stanley, 'Introduction', in *Future Primitive: The New Ectopias*, ed. Kim Stanley Robinson (New York: Tor, 1994), 9–11.
Robinson, Kim Stanley, *Blue Mars* (New York: Bantam, 1996).
Robinson, Kim Stanley, *Antarctica* (London: Harper Voyager, 1997).
Robinson, Kim Stanley, *Green Mars* (London: Harper Voyager, 2001).
Robinson, Kim Stanley, *Red Mars* (London: Harper Voyager, 2009).
Robinson, Kim Stanley, 'Remarks upon Utopia in the Age of Climate Change', *Utopian Studies* 27, no. 1 (2016): 2–15.
Stapledon, Olaf, *Last and First Men* (London: Penguin, 1930).
Tchaikovsky, Adrian, *Children of Time* (London: Pan Macmillan, 2016).
Teilhard de Chardin, Pierre, *Christianity and Evolution*, trans. by René Hague (San Diego: Harcourt Brace and Company, 1969).Turner, Frederick, *Genesis: An Epic Poem* (Dallas: Saybrook, 1988).
Whitehead, Alfred North, *The Concept of Nature* (Cambridge: Cambridge University Press, 1920).
Whitehead, Alfred North, *Science and the Modern World* (New York: Macmillan, 1925).
Whitehead, Alfred North, *Process and Reality: An Essay in Cosmology* (New York: Macmillan, 1929).

Chapter 9

A PHILOSOPHICAL OUTLOOK ON POTENTIAL CONFLICTS BETWEEN PLANETARY PROTECTION, ASTROBIOLOGY AND COMMERCIAL USE OF SPACE

Erik Persson
Lund University

Introduction

The aim of this chapter is to use philosophy and, in particular ethical theory, to identify and explore some potential conflicts between planetary protection, astrobiology and commercial use of space.

According to NASA's astrobiology roadmap, astrobiology is a research field that addresses the following questions: 'How does life begin and evolve, does life exist elsewhere in the universe, and what is the future of life on Earth and beyond?' (Des Marais et al. 2003; 2008). I will, however, concentrate on the most well-known and most curious part, namely the search for life outside our planet. This might sound a lot like science fiction, and the truth is of course, that we have still not found any extra terrestrial life (at least, not at the time of writing). The search for extraterrestrial life is not science fiction, however. It is, in fact, a very serious scientific endeavour that the scientific community judge to have a high probability of success in the not too distant future. The fact that we have still to find extraterrestrial life also makes it an unusually suitable subject for ethical discussions. Contrary to what is typically the case on our own Earth, the fact that we have not yet found any extraterrestrial life gives ethics a chance to influence our decisions from the beginning, instead of just pointing out where we went wrong *ex-post-facto*.

Commercial use of space might seem just as much like science fiction as the discovery of extraterrestrial life, but the fact is, it is already happening (see Beavin 2008; Beery 2012; Giacalone 2008; Hubbard et al. 2013; Lewicki et al. 2013; Peeters 2003), and the road is being paved for a much more intensified commercial use of space. This development is interesting from many different perspectives, and it evokes many complex ethical questions.

Planetary protection, the way the term is used today within the astrobiology community, is essentially a technical term for measures aimed at avoiding biological contamination, either of extraterrestrial environments (forward contamination) or of our own Earth (back contamination) resulting from human space activities. The legal basis for planetary protection can be found in the *Outer Space Treaty* from 1967, stating in Article IX that 'Parties to the Treaty shall pursue studies of outer space, including the Moon and other celestial bodies, and conduct exploration of them so as to avoid their harmful contamination and also adverse changes in the environment of the Earth resulting from the introduction of extraterrestrial matter and, where necessary, shall adopt appropriate measures for this purpose' (UN General Assembly 1966).

As is usually the case with international treaties, it states over arching principles rather than precise goals or instructions. We can therefore not conclude exactly what 'harmful contamination' means, or what needs to be done to avoid it, based solely on the treaty. The responsibility for setting up more detailed guidelines regarding planetary protection has fallen on the Committee for Space Research (COSPAR), under the International Council for Science (ICSU). The first Planetary Protection Policy developed by COSPAR was published in 1964 (COSPAR 1964). The latest planetary protection policy, put together by COSPAR's Panel for Planetary Protection (PPP), was published in 2015 (Kminek and Rummel 2015). It specifies which degree of protection is needed for different types of activities (probe, lander, flyby, orbit) on different bodies (planets, moons, asteroids) depending on their estimated interest to astrobiology. Mars and Europa are particularly interesting to astrobiology, and there are therefore special requirements for flyby, orbiting and landing on, as well as sample return missions from these bodies. There are also special requirements for certain regions, so-called 'Special Regions' on Mars, where it is thought more likely that earth organisms could survive (Kminek and Rummel 2015).

Strictly speaking, the policies are recommendations and not legally binding (though the Outer Space Treaty brings in those countries that have ratified the treaty). In practice, the COSPAR guidelines are respected by the space agencies of the major space-faring nations, though they themselves typically specify the COSPAR recommendations even further to fit individual missions (see, for instance, *https://planetaryprotection.nasa.gov/overview*, accessed 9/2/2017).

When put in a wider context, this includes several intricate philosophical conundrums. In fact, both forward and back contamination have interesting ethical implications, but since my focus here is on extraterrestrial life, I will only discuss forward contamination in this chapter.

Planetary protection, astrobiology and commercial activities in space are three human endeavours that are in some ways very different, but that also have two major aspects in common: They are all related to human activities in space, and they all have the potential to affect living organisms in ways that need ethical deliberation. When these three activities are combined, we get a whole set of new, and, from a philosophical point of view, exciting questions. The main aim of this chapter is to point at, explore and, to some extent, suggest ways to handle some of these questions.

Astrobiology and Planetary Protection

Planetary protection is extremely important for astrobiology. When looking for extraterrestrial life, it is obviously crucial to avoid false positives due to contamination by earth life. If and when we ever find extraterrestrial life, it is also very important to be able to study it in its pristine form without contamination from earth life, and of course, most important of all, that the indigenous life is not driven to extinction by invasive earth organisms before it is even discovered. When sending unmanned as well as manned missions to other worlds, it is therefore vital to avoid contamination of that world. Considering how important planetary protection is to astrobiological research, it is easy to believe that there is no conflict between the two. This is not completely true, however. Even though efforts to avoid contamination are quite sophisticated and considerable work is put into the process, we also know that complete sterilization of humans and machines is impossible (see Kim et al. 2013; La Duc et al. 2003; 2007; Moissl et al. 2008; Mahnert et al. 2015; Newcombe et al. 2005; Schuerger and Nicholson 2016; Stieglmeier et al. 2009; Tepfer and Leach 2017; Dartnell et al. 2010 for some examples of the challenges involved). We can obviously never totally sterilize humans. If we did, the humans would die too. We can go somewhat further with machines but essentially, the same is true for them. The electronics in a rover is more sensitive than at least some earth microbes. We also know now that what kills some bacteria allows others to flourish. Interestingly, our efforts to sterilize exert a selective pressure in favour of strands that are resistant to these efforts. Planetary protection is thus something that comes in degrees and it will always be possible to do more or less than we already do. This means that we must make decisions about exactly how far we want to go. It would be easy to say that the more the better, but the only absolutely safe way would be not to send anything to any other world. As long as we want to go on exploring, in space as well as on Earth, we have to find a compromise between the importance of discovery and the risks we impose on the study objects in the process.

Another aspect to consider is that planetary protection, like everything else, comes at a cost in terms of time, work and money that could be used on other things, also worthwhile. The experience from the Viking landers, that were the most thoroughly decontaminated landers so far, confirms that decontamination can be prohibitively expensive (Newcombe et al. 2005). Schon (2009) opposes a higher standard of sterilization because it would mean higher costs, and some astrobiology researchers complain that even today's standards are too high, or as one researcher put it: 'Over the top' (private communication).

These facts taken together show that protecting the life we are looking for and want to study is important, and so is shielding experiments from false positives. It also shows, however, that planetary protection comes with certain costs and that the search in itself also poses a very real danger to the extraterrestrial life we are looking for. If we still want to be able to look for and study extraterrestrial life, we need to find a balance between our efforts to find and study it and our efforts to protect it.

COSPAR has provided an explicit answer as to where that balance can be struck, namely at the degree of planetary protection needed to give science enough time to find and study possible life (see, for instance, Sagan and Coleman 1965; Kminek and Rummel 2015). They have thus chosen a degree of planetary protection that has a strictly defined purpose with clear limitations in both degree and time.

The aim of planetary protection, as it is formulated and applied by COSPAR, is sometimes stated as being a matter of 'protecting the science' rather than protecting extraterrestrial life as such. A more philosophically stringent way of putting it would be to say that extraterrestrial life is protected only because of its epistemic instrumental value. Here, 'epistemic value' refers to value that has to do with knowledge or the production of knowledge, while 'instrumental value' refers to the value something has because of its ability to promote something else that has value (in this case, knowledge). Taken together, referring to the value of extraterrestrial life as epistemic instrumental value, means that extraterrestrial life has value as a source of knowledge, that is, as objects for study.

The question we should ask, however, is: Could it be possible that extraterrestrial life in addition to its value as study objects, also have other values? If it is, then we also need to ask: What practical consequences does that have for the relation between astrobiology and planetary protection?

It is conceivable that extraterrestrial life can have other values, but that these values will be sufficiently similar to the epistemic values, or be weaker or less vulnerable than the epistemic values, and thus confirm that the present guidelines for planetary protection are sufficient to also protect these values. Another possibility is that they have values that are even more important or more vulnerable, and that we will therefore need even stricter guidelines to achieve an even lower probability of contamination, compared with today. The most extreme scenario will, of course, be that we have to set the maximum acceptable probability of contamination so low that sending spacecraft from Earth to a potentially inhabited world will in practice (and maybe even in principle) be ruled out. This would be a very sad conclusion and it is not something I see as very plausible, but it at least needs to be discussed. Another and much more plausible possibility is that we will have to abolish time limits. If extraterrestrial life has value in any other way than as study objects, it is very hard to justify why they should only be protected for the time it takes to study them. Yet another possibility is that there will be restrictions on what we are allowed to do in order to find extraterrestrial life, and what we are allowed to do to them when we find them. Will we, for instance, be allowed to use destructive detection methods as was the case with the Viking experiments? (see Horowitz et al. 1972; Hubbard 1976; Klein 1978). Will we be allowed to bring them back to Earth to study them, and will we be allowed to dissect or perform other intrusive experiments on them?

What is important to remember is that these questions are not science questions. They are value questions. They cannot be settled by science alone, and this is where philosophy enters the picture.

Let us summarize our options:

1. Extraterrestrial life has no other value than its epistemic value. In this case, all we have to do is to figure out the exact level of protection needed to protect their value as study objects as long as is needed to study them thoroughly, and then figure out how to achieve that level of protection. These tasks are by no means easy but at least they are scientific and technological/methodological tasks. If this is our conclusion, there will be no more need for philosophy.
2. Extraterrestrial life has at least one other kind of value but it is a kind of value that does not warrant any extra protection in addition to the planetary protection guidelines. The practical implications of this conclusion are the same as for 1.
3. Extraterrestrial life has at least one other kind of value that is sufficiently strong to warrant protection but it is a kind of value that does not warrant any extra protection in addition to the planetary protection guidelines. It does, however, warrant limitations as to what we are allowed to do to detect it, study it or utilize it in some other way. In this case, the present level of planetary protection is probably sufficient, but other restrictions need to be set regarding detection methods and what we will be allowed to do to this life once we find it. The latter concerns will not be of direct concern for planetary protection the way it is presently defined, but it has to be handled some way. One way of handling it would be to widen the present interpretation of the term 'planetary protection', which would call for a reinterpretation of the Outer Space Treaty. Another way would be to set up an additional treaty to handle these questions. Both these options would require a body that can handle the details in a similar way to what COSPAR is doing today. This could be done by extending the mandate of COSPAR to also deal with these issues, or by instantiating a completely new organization for this purpose, or to give this mandate to some other existing organization.
4. Extraterrestrial life has also at least one other kind of value that is strong enough (it could be stronger, weaker or equal to its value as study objects) to warrant protection but that has no end point in time. In this case, we need to abolish the time limit for planetary protection, which will, in turn, affect how we calculate the maximum acceptable probability of contamination for individual missions.
5. Extraterrestrial life has at least one other value that is strong enough or a number of other values that together are strong enough (if one believes in the accumulation of values) to make it trump all other values. In this case, it will not be a matter of finding a balance. We simply have to stop doing anything that could interfere with this life.

Deciding which one of these options is the correct one is too big a question to be answered in this chapter. Since it is a value question, it is not subject to the same process of shrinking the room for interpretations that empirical findings provide for scientific theories. We will thus probably never reach a situation where we can all agree on one particular option. This does not mean the discussion is meaningless, however. On the contrary, an ongoing discussion that can shape the public policy regarding where to strike the balance between exploration and planetary protection, and between these two and a wider societal and ethical perspective, will be of crucial importance in order to make the policy decisions as ethically justifiable as possible. Later in this chapter, we will discuss some alternative ways in which extraterrestrial life can have value and what implications that may have. Before we get to that, it is time to introduce another complication, however.

Adding Commercial Space Use

In the previous section, we found that the relation between astrobiology and planetary protection is more complicated than one might expect at first sight. In this section, we will show how things become even more complicated when we add commercial space use to the formula.

The question of legal rights for commercial initiatives in space is very important. Anyone who invests major sums of money in an activity wants to make sure that they will not lose their investments because it turns out afterwards that they did not have the right to the resources they extract. It is also important to have laws in place that regulate the relations between different companies from different countries, bound by different laws in their home countries, but trying to coexist in space.

There are also questions of coexisting with other players with other agendas, such as the scientific community. It can be expected that space research and commercial space projects will prove to be very useful for each other. Space research is necessary for commercial companies to make sound business decisions about their activities in space. Commercial space companies will also be, and to some extent already are, useful for science. They are, for example, already providing launch capacities. This service will probably be the most valuable commercial contribution to space science, at least in the short term. Many hope that private launch solutions will be more cost efficient than those provided by major national and international space agencies such as NASA, ESA and Roscosmos. In the future, space mining may also provide access to resources, such as fuel and construction materials in space, which will make it possible for the scientific community to reach further with more ease and less money, and to build and operate observatories in space, etc.

In addition to the many opportunities for cooperation that can be expected, there will, however, also be points of conflict. Will planetary protection be such a point of conflict? There are not yet any commercial activities on Mars or any other extraterrestrial body, and the most immediate plans for space tourism are only about sending people into Low Earth Orbit, or even just about making a short 'jump' into space to let the passengers experience a short moment of weightlessness. Likewise, the immediate plans when it comes to space mining are about asteroids, which are considered by COSPAR to be among the least probable places to find life in our solar system (Kminek and Rummel 2015). Even so, at least one company has foreseen that their operations might be controversial. They have therefore produced a web page entitled 'Asteroids are cold, dead worlds', seemingly aimed at setting the reader to rest that asteroid mining really does not have any objectionable environmental impact on the target bodies, and also that mining and using resources from asteroids will even be a good thing for the earth environment (https://deepspaceindustries.com/asteroids-are-cold-dead-worlds/ accessed 03/01/2017).

It is not a very controversial assumption, however, that when the time comes to exploit the resources on a potentially inhabited world, or to establish it as a tourist destination, we will see conflicts between those who think that we have looked for life long enough on the world in question and that it is now time to give the green light for development, and those who think that there is still a chance there might be life that we should not endanger.

It all depends on the values at stake, of course. If we assume that extraterrestrial life only has value as a study object, it might be very difficult to resist exploitation even if it potentially puts indigenous life at risk. Economic value is privileged by our society, and when the value of knowledge as such is set against economic value, the former usually loses.

An alternative possibility is that extraterrestrial life, in addition to its value as a study object, also has economic value. Just like with the value as a study object, this is also a form of instrumental value: Something has economic value because it can generate money. Can extraterrestrial life have instrumental value in this way, and what does it mean for the relations between astrobiology, commercial interests and planetary protection if it does?

Cockell (2011a) mentions bioengineering as an example. He distinguishes between the economic value of extraterrestrial life that is related to us (spread through the solar system according to the so-called panspermia theory) compared to if it is not related to us. This distinction makes good sense. It is easy to imagine that a microbe that is genetically well adapted to life on another world might contain adaptations that we will want to insert into earth bacteria and use for different purposes. In such cases, it clearly makes things easier if they are genetically related. If they are, we will be able to transfer the properties in question to earth life by transferring the relevant genes from extraterrestrial microbes to earth microbes. Interesting properties could, for instance, be the ability to survive high doses of radiation, which could be useful if we, for example, want to engineer microbes to do work inside a nuclear reactor or disaster area, or aboard a spaceship travelling between Earth and Mars. An ability to extract energy from the Sun in a very efficient way would be another example of a useful property that might be found in microbial life on worlds further from the Sun. In addition, it is also easy to imagine that life that is already adapted to life on another world but is also compatible with earth life will be useful if we want to transfer traits from earth microbes to extraterrestrial microbes in order to tailor make microbes that future space-dwellers can use to produce anything from food to fuel in their new habitat.

When it comes to extraterrestrial life that is not genetically compatible with earth life, it would be more difficult to transfer properties between them and earth life in either direction, even if not in principle impossible (maybe we will be able to extract the relevant information from their equivalent to a genome and translate it into the language of DNA). Even if extraterrestrial life cannot be genetically mixed with earth life, they could be used for bioengineering, however, without involving earth life in the process.

If we find extraterrestrial life that has value as a resource for bioengineering, it could mean that the aims of science and business actually converge when it comes to planetary protection. In order to use extraterrestrial microbes for engineering, they will need to be protected so they do not go extinct, but they will also need to be found and studied. This indicates that the degree of protection suitable for extraterrestrial life as resources for bioengineering could be very similar to the degree of protection suitable for extraterrestrial life in their role as study objects. There is one difference, however, namely that the limit for how long they need protection may not be the same. If we want to use them for bioengineering, they need to be protected beyond the time it takes to study them. On the other hand, the time it takes to study them enough to map their economically useful properties may be shorter than the time it takes to study them thoroughly from a scientific perspective. The value of extraterrestrial life as study objects and as resources for bioengineering may thus make different demands on the time-span during which they have to be protected, and therefore, also on the degree of protection. Which type of value that will be more demanding is not possible to say before we have actually found them.

Another way in which extraterrestrial life can have economic value is in the form of entertainment. This too demands some degree of planetary protection, though the demands will be different compared to the demands from bioengineering. We can easily imagine that the interest in seeing extraterrestrial life with one's own eyes will be immense. This value can be exploited in two different ways, either by bringing people to their world, or by bringing the extraterrestrials to us. Both these alternatives will put extra pressure on planetary protection. We have mentioned that it is impossible to completely sterilize humans. This means that bringing human tourists to see extraterrestrial life in situ carries a substantial probability of being destructive. If they are compatible with earth life, it might also imply a risk for the tourists. Bringing the extraterrestrials to Earth will for obvious reasons also imply risks, in this case for us and other earth life. Sterilization will not be an option in this case (we do not want to kill the life we have brought such a long way to see).

All in all, we can conclude that if we look at the question of planetary protection from the perspective of science or from the perspective of business, and even when we consider different business perspectives, the result may differ considerably. Considering the expected increase in commercial space use, this obviously complicates things for science as well as for planetary protection.

Does Extra terrestrial Life have Instrumental Value Other than Economic and Epistemic Value?

Are there any other ways in which extraterrestrial life can have instrumental value than the two forms of value discussed above? The answer seems to largely depend on which properties they turn out to have. Different properties mean different uses and thus different bases for instrumental value. We can be sure that they will have epistemic value, and it seems plausible that they will also have economic value in some form, if not for bioengineering, at least as entertainment. Saying anything else about their potential instrumental value before we have found them, and thus before we know their properties in any detail, is difficult. What we can say, is that whatever kind and degree of instrumental value they have, that value has to compete with other values. If we find other resources that can be extracted from their world and they somehow will be in the way of that, it will be hard to fight for their protection, just as it has been and still is with life on our own planet. If we can establish that the extraterrestrial life forms also have non-instrumental value, it would definitely change the game. It might still not be enough to demand a higher degree of planetary protection or to say that there are certain things we will not be allowed to do to them, but it might, and therefore, I think it is our duty to investigate the possibility.

Does Extra terrestrial Life Have Non-instrumental Value

Can we imagine extraterrestrial life having value beyond their instrumental value? This question often takes the form of a question whether extraterrestrial life has *intrinsic value* (see, for instance, Cockell 2011a, b; Cockell and Horneck 2006). Since different authors tend to use the term 'intrinsic value' in different ways, this way of formulating the question is hopelessly opaque, however. Attempts to untangle the different meanings of the term (O'Neil 1997; O'Neil 1992; Persson 2008; Rabinowicz and Rønnow-Rasmussen 1999) have revealed at least four different meanings:

1. Value as an end in itself, independently of what other values it can promote (also known as 'end value', 'final value' or just 'non-instrumental value'),
2. value that is inherent in the phenomenon in question and thus independent of its relations to other phenomena (also known as 'inherent value' or 'non-relational value'),
3. value that is independent of whether it is valued by someone (also known as 'objective value'),
4. moral standing, which means that the phenomenon in question has interests of its own that need to be considered by anyone who has the ambition of acting morally. Strictly speaking, the question of moral standing is not about the *value of* the phenomenon in question, but rather about whether things can have *value to it*.

For the purpose of this chapter, we only need to pursue two of the different meanings of 'intrinsic value', namely, the first ('end value'), and the last ('moral standing'). We thus need to ask: Does extraterrestrial life have end value, and/or does it have moral standing?

Does Extraterrestrial Life Have End Value?

Before we start investigating whether extraterrestrial life can have end value we need to be aware that instrumental value and end value are not mutually exclusive. It is perfectly possible to have both. Something can have value as an end in itself, and also be a useful means to something else. A watch can, for instance, be very useful for keeping time but also be a precious heirloom. A trusty old car that has been with its owner through many adventures can be convenient for getting from A to B and thus have instrumental value, but when it is stolen, it can still evoke a feeling in the owner that she has lost something more dear to her than just a means of transportation. It can have end value to her in addition to its instrumental value. The fact that we have found several ways in which extraterrestrial life can have instrumental value does not therefore compel us to give either a positive or a negative answer to whether it has end value.

So, can extraterrestrial life have end value? The simple answer to this question is: Yes, anything can be valued as an end in itself by anyone! This is true. We might, however, also want to ask whether it is reasonable to value extraterrestrials as ends in themselves. The answer to that question is also yes, for several reasons. One reason is that life as such seems to be one of these things that most people value for its own sake independently of its instrumental use. A large part of the motivation behind the environmental movement can, for instance, be attributed to this intuition (Persson 2008). The fact that we talk about extraterrestrial life can add to this value, at least based on the huge interest among the general public for the question of whether extraterrestrial life exists, an interest that is supposedly not only based on the value of extraterrestrial life as a study object or other instrumental values it might bring. In addition, the circumstances during which we find extraterrestrial life, as well as what kind of life we find, might add to its end value. Being first seems to be a property that is a strong basis for end value for many people. The first extraterrestrial life we find will therefore surely have end value in its capacity of being the first discovered extraterrestrial life. Uniqueness is another property that is a basis for end value to many people (Persson 2017). As long as we only find one instance, it will therefore also have end value because it is unique. If we find extraterrestrial life on more than one world, it will not be unique in this way, but since life on one world will probably differ considerably from life on another world, it can be safely assumed that they will all have properties that make them unique in one way or another (Persson 2017). In fact, this is true even of the different forms of life we already know on our own planet today. All life on Earth is related but has evolved in many different ways, resulting in an almost endless number of unique solutions. Considering the differences in physical conditions between different worlds, we have reason to believe that this will be the case to an even higher degree if we look at life on different worlds, and this as such seems like a good reason to attach end value to extraterrestrial life.

It can also be expected that the extent to which the life we find has properties that make it interesting or exciting will play a role. More complex life will probably be considered more valuable than less complex life independently of whether it makes the life in question more instrumentally useful or not. In the same way, life that expresses a more complex behaviour or is aesthetically more pleasing to us will probably be considered more valuable than life with a less complex behaviour or that is less aesthetically pleasing. Another property that, with a high degree of certainty, can be said to add end value to a life form, from our perspective, is its ability to communicate with us. The more we can interact with it, the more exciting will it be, which in turn will make us value it higher as an end in itself, no matter whether it makes it more or less useful to us.

To what degree a certain life form has end value thus depends on its particular properties, but we can also conclude that all extraterrestrial life will have some degree of end value based on the fact that it is life, and that it is extraterrestrial.

Does Extraterrestrial Life Have Moral Standing?

Having moral standing, or being a moral object as it is also called, is not the same as having end value. Strictly speaking, it is not about the value of the object at all. To say that you have moral standing means rather that things have (positive or negative) value to you, and that is something that has to be considered by others.

To answer the question of whether extraterrestrial life has moral standing, we need to consider two things: What kind of life are we talking about (which properties will it have), and what are the criteria for having moral standing (which properties are necessary for having moral standing)?

Obviously, we do not know what kind of life we will find, and a philosopher's job (in spite of popular belief) is not to speculate about what is at least in principle an empirical question, but rather to analyse and scrutinize the non-empirical questions, so let us keep an open mind regarding the question of what kind of life we might find, and consider what different theories about moral standing have to say about the moral standing of different kinds of life, but with the proviso that planetary protection is, given current technology, only applicable to activities in our own solar system, and that there seems to be a rather strong belief among astrobiologists that as long as we talk about our own solar system, we will probably not find anything very complex. I will, therefore, put extra focus on what the different theories about moral standing have to say about the simplest kinds of microbial life.

The question of what it takes to have moral standing is intensely discussed in environmental ethics on our own planet, and the answers are typically divided into the following categories:

1. Anthropocentrism – all and only humans have moral standing (Carruthers 1994; Kant 1998; Smith 2009).
2. Biocentrism – all and only living beings have moral standing (Schweitzer 1976; Goodpaster 1978; Cockell 2005; 2011a).
3. Sentientism – all and only sentient entities have moral standing (see, for instance, Bernstein 1998; Clark 1977; de Grazia 1996; Helm 2002; Jamieson 1998; Levine 1997; O'Neil 1997; Regan 1983; 2001; Singer 1993; 1995).
4. Ecocentrism – Species and biomes have moral standing in addition to living beings (see, for instance, Callicott 1980; 1985; 1987; 1992; 1999; Johnson 1991; 1992; Leopold 1970; Plumwood 1991; Rolston 1986; 1987; 1988; 1994; 1999).

Here I will briefly discuss what these theories imply for extraterrestrial life (for an extended discussion see Persson 2012).

Anthropocentrism

If we want to apply this theory to the question of whether extraterrestrial life has moral standing, we need to distinguish both between different kinds of extraterrestrial life and between different versions of the theory. If we adhere to the simplest version of the theory, namely that the criterion for having moral standing is simply to belong to the species *Homo sapiens*, then we can be quite sure that no extraterrestrial life will ever have moral standing. Even in the seemingly unlikely event that we find extraterrestrial life forms that look just like us and share most of our features, they will not be the same species as we, and thus not human in this very strict sense.

The basis for anthropocentrism is usually not species membership as such, however. Instead, the focus is typically on some property that, on this planet, is supposed to only be possessed by humans. Examples of such properties could be intelligence or language skills on a level that is beyond what is found among other species on our planet.

Not even the most optimistic astrobiologists believe that we will find extraterrestrial life with any humanlike properties in our solar system. This means that if we accept anthropocentrism, we do not have to worry about considering the moral status of extraterrestrial life in connection with planetary protection, as long as we, as is the case here, only talk about forward contamination, and as long as we stay within our solar system.[1]

1. See Lehmann Imfeld, this volume, for a discussion of the ethical implications of anthropocentrism in the terraforming of Mars.

If we extend our discussion outside the solar system, we do not know what we might find and it is at least not impossible that we will find life forms that share some of our properties. Applying anthropocentrism on our relations with them, the result may be mostly detrimental to us, however. If we base moral status on the property of being human in a strictly biological sense, how can we argue against any possible alien invaders, who by the same logic claim that they do not have any moral duties to consider our interests since we do not belong to their species?

If we instead use, for instance, intelligence as the basis for moral standing and they are much more intelligent than we are, maybe so much more that our level of intelligence, in their eyes, is not significantly different from the intelligence level among the earth animals that we use for our own purposes today, what then? Should we just succumb to our new role as slaves or food for our new overlords or do we have any moral basis from which we can argue that their treatment of us is wrong? It might be hard to do that and at the same time maintain that we have the right to treat other, from our perspective, lower species, the way we do today.

Biocentrism

At first sight, it may seem as if biocentrism would demand a very high standard of planetary protection, including sterilization of all equipment aimed at, or in risk of landing at potentially inhabited worlds. If we think one step further, however, we realize that what is sterilization other than the systematic mass killing of earth microbes? It would, from a biocentric point of view, be very hard to motivate planetary protection measures that imply the killing of countless numbers of actual Earth life in order to protect merely possible extraterrestrial life based on the moral standing of all life. It thus seems that the conclusion has to be that we have to completely abstain from a central component of planetary protection, that is, sterilization. Instead, we have to decide whether to go without any sterilization measures or not to go there at all. This decision presumably has to be based on how probable we think it is that the target world is inhabited and how probable we think it is that it is inhabitable for earth microbes.

The latter alternative (not going at all), is perhaps too restrictive even from a biocentric perspective, however. Granting moral standing for extraterrestrial life does not necessarily mean that we are not allowed to do anything on an inhabited planet. How prohibitive an ethical theory is depends on how we answer all the ethical questions, not just how we answer the question of what it takes to have moral standing. If we, for instance, subscribe to a utilitarian view of ethics, the interests of all moral objects have to be considered and weighed into the equation. The outcome depends on the whole equation containing all affected interests according to their relative strength.

In either case, if we decide that it is acceptable to go to their world, we will have certain obligations to protect the extraterrestrials. Even if this will not take the form of sterilization of spacecraft, it can take other forms that will affect the study of these life forms. It may, for instance, imply that we have to abstain from destructive experiments (that is, experiments in which the study objects are killed or injured) and from taking them from their own environment and bring them back to Earth for closer examination.

Sentientism

As long as we stay within our solar system, sentientism probably has the same immediate consequence for planetary protection as anthropocentrism, namely, that the life we expect that we might find in will not have any moral standing. There is, of course, some difference in probability. We do not know what hides under the ice of Europa, for instance. Even though the probability that it will be sentient is very low, it is not as low as the extremely low probability that it is human like.

If we look beyond our present capabilities and think about the implications of sentientism when we achieve the technological ability to detect biosignatures outside our solar system, the most obvious challenge will be to distinguish between sentient but not intelligent life on one hand, and non-sentient life on the other. This is in some cases a difficult challenge even on our own planet. When we encounter life that might be radically different from life as we know it and has evolved under radically different circumstances, how do we then decide whether it is sentient? This is a question that we cannot yet answer, though that does not of course mean that we should abstain from using sentience as the decisive criterion for moral standing if this is the criterion that makes the most sense from an ethical perspective (which seems like a very reasonable assumption). It means, however, that we have to be extremely careful when the time comes to make a decision about whether a particular extraterrestrial is sentient or not.

Ecocentrism

If we accept ecocentrism, what implications would it have for planetary protection? One implication that is potentially very important, is that we not only will have to consider any individuals that we find, but also should avoid altering competitive relationships between species, or otherwise interfere in their relationships. This means, for instance, not drastically changing the environment on their world.

On the other hand, killing some instances of individual extraterrestrial life would not be a big problem for the ecocentrist as long we do not threaten the whole species or the environment on the planet. This is potentially very important for the relationship between planetary protection and astrobiology. A theory that allows you to kill off a few, or a few hundred thousand individuals, as long as you do not threaten the species or the function of ecosystems, seems perfect for justifying both decontamination, where uncountable numbers of presumably very common earth microbes are killed off, and research that includes the killing of a number of individual extraterrestrial microbes in the process. On the other hand, though both anthropocentrism and sentientism deny moral status to individual microbes and for species, they accept that species can have value (though not moral standing). Both these theories would thus provide exactly the same allowances as ecocentrism as long as we talk about microbial life. That is, from an anthropocentric or sentientistic point of view, it would also be okay to kill individual microbes as long as we do not threaten the species, given that the species have the kind of instrumental or non-instrumental value discussed above.

As long as we talk about microbial life and are only concerned about forward contamination, anthropocentrism and sentientism thus have the same practical implications as ecocentrism. If we find sentient or even intelligent life, on the other hand, the practical implications would differ considerably. In these cases, sentientism and anthropocentrism would set restrictions for what we are allowed to do with the individual organisms that ecocentrism would not. Ecocentrism does accept some level of moral status for sentient non-human individuals but it is much lower than the moral status of the species. The status of human or human like life according to ecocentrism differs between different versions of the theory. According to some versions of ecocentrism (Rolston 1986; 1987; 1988; 1994; 1999), human individual life has a special status that puts it above other individuals and gives us some unique rights in relation to other species and ecosystems. Exactly which implications this has when we talk about human interest in space exploration as well as in commercial activities is not really worked out by its proponents, though a fair guess would be that we would have relatively generous rights to explore and exploit individual extraterrestrial life as long as it does not threaten entire species or their environments.

According to another version of ecocentrism called *land ethics*, extraterrestrial life is explicitly excluded from the realm of moral objects (Callicott 1992). If we find life on another planet, we will for the first time have to deal with life that we are not related to, even distantly (provided of course that life has not been transferred between Earth and this other world in either direction). This means in turn that according to land ethics, extraterrestrial life cannot be included in our moral circle.

A general point to consider when discussing ecocentrism in relation to space exploration and exploitation is that ecocentrists usually do not show much concern for non-biological systems, which means that a world populated by intelligent robots will not have any moral standing according to this theory, even if they are conscious and/or intelligent. A maybe more immediate concern is that it is also unclear what happens to the moral standing of an extraterrestrial organism that is taken from its natural habitat and moved to Earth. When that happens, the organism will no longer fulfil its role in the species and ecosystem to which it originally belonged. Advocates of ecocentrism tend to consider domesticated animals and animals in zoos more as artefacts than as 'real' animals (Johnson 1991; Rolston 1988; 1994). It is possible that extraterrestrial life brought back to Earth will be considered in the same way and thus lose most of their moral status.

Summary and Conclusions

The relation between astrobiology, commercial space activities and planetary protection evokes a number of philosophical questions, not least value questions. Value questions cannot be handled by science alone but they are, nonetheless, of utmost importance to science as well as for commercial space initiatives. In this chapter, I have identified and tried to throw some philosophical light on some of these questions.

The most basic value question in the intersection of astrobiology, commercial space activity and planetary protection is, of course, which values are at stake? It turns out that the value that solely motivates planetary protection today is the value of extraterrestrial life as study objects, that is, the instrumental epistemic value of extraterrestrial life. It was also found, however, that this is not the only possible value at stake. For the commercial actors, the economic value of extraterrestrial life is probably of greater importance, which evokes another value question, namely, how the epistemic and economic values of extraterrestrial life relate to each other. It was found that in order to utilize extraterrestrial life for economic purposes, it has to be studied, so in that respect, the two values seem to be compatible. Extracting the economic value of extraterrestrial life can, on the other hand, interfere with the study of them, while the assumption that the study has an end point in time might mean that the economic value of extraterrestrial life calls for continued protection beyond the time needed to study it. Another point of conflict that is potentially more serious is that other things such as non-living resources may have a higher economic value than the extraterrestrial organisms. This can be expected to lead to conflicts over how long the search for, and subsequent study of extraterrestrial life, with accompanying high standards for planetary protection should be allowed to go on.

It is also plausible that extra terrestrial life has other forms of value, instrumental value as well as end value. Exactly which values and how valuable will depend on the specific properties of the extraterrestrial life. All extraterrestrial life will share a minimum end value, supervening on it being alive and extraterrestrial, which includes an aspect of uniqueness. This, in turn, should be enough to motivate abolishing the time limit on planetary protection set by its value as study objects.

Another important question that has to be looked into is whether extraterrestrial life, in addition to any value it might have, also has moral standing. There are different theories about what it takes to have moral standing, which means the answer to this question depends on a combination of what kind of life we find and which theory we use.

How should we handle the conclusion that it is not just possible, but very reasonable to assign other values to extraterrestrial life, including economic value and other instrumental values, as well as end value, in addition to its epistemic value?

One way forward would be to widen the basis of planetary protection to also include these other values. A problem in connection with this suggestion is that planetary protection, the way it is done today, is based on a well-established international treaty, the so-called Outer Space Treaty. This treaty is the basis for national legislation regarding not just planetary protection, but also many other space-related questions (UN General Assembly 1966). Making changes in the value basis behind planetary protection might violate this treaty and subsequently national and international legislation. An alternative way of handling the value issues identified in this chapter would be to complement the Outer Space Treaty with another treaty – an interplanetary environmental protection treaty.

References

Beavin, Michael, 'Office of Space Commercialization Promotes U.S. Space Industry', *Earth System Monitor* 16 (2008): 7.

Beery, J., 'State, Capital and Spaceships: A Terrestrial Geography of Space Tourism', *Geoforum* 43 (2012): 25–34.

Bernstein, Mark, 'Well-being', *Am Philos Q* 35 (1998): 29–55.

Callicott, J. Baird, 'Animal Liberation: A Triangular Affair', *Environ Ethics* 2 (1980): 311–38.

Callicott, J. Baird, 'Book Review of *The Case for Animal Rights*', *Environ Ethics* 7 (1985): 365–72.

Callicott, J. Baird, 'The Conceptual Foundations of the Land Ethic', in *Companion to a Sand County Almanac—Interpretive Critical Essays*, ed. J. B. Callicott (Madison, WI: The University of Wisconsin Press, 1987), 86–217.

Callicott, J. Baird, 'Moral Considerability and Extraterrestrial Life', in *The Animal Rights/ Environmental Ethics Debate*, ed. E. C. Hargrove (Albany: State University of New York Press, 1992), 137–50.

Callicott, J. Baird, *Beyond the Land Ethic* (Albany: State University of New York Press, 1999).

Carruthers, Peter, *The Animals Issue* (Cambridge: Cambridge University Press, 1994).

Clark, Stephen R. L., *The Moral Status of Animals* (Oxford: Clarendon Press, 1977).

Cockell, Charles S., 'The Value of Microorganisms', *Environ Ethics* 27 (2005): 375–90.

Cockell, Charles S., and Gerda Horneck (2006) 'Planetary Parks – Formulating a Wilderness Policy for Planetary Bodies', *Space Policy* 22 (2006): 256–61.

Cockell, Charles S., 'Ethics and Extraterrestrial Life', in *Humans in Outer Space – Interdisciplinary Perspectives*, ed. Nina-Louisa Remuss, Kai-Uwe Schrogl, Jean-Claude Worms and Ulrike Landfester (New York: Springer, 2011a), 80–101.

Cockell, Charles S., 'Microbial Rights?' *SMBO Reports* 12 (2011b): 181.

COSPAR, 'COSPAR Resolution 26.5', *COSPAR Information Bulletin* 20 (1964): 25–6.

Dartnell, Lewis R., Stephanie J. Hunter, Keith V. Lovell, Andrew J. Coates and John M. Ward, 'Low-Temperature Ionizing Radiation Resistance of Deinoccocus Radiodurants and Antarctic Dry Valley Bacteria', *Astrobiology* 10 (2010): 717–32.

de Grazia, David, *Taking Animals Seriously* (Cambridge: Cambridge University Press, 1996).

Des Marais, David J. et al., 'The NASA Astrobiology Roadmap', *Astrobiology* 3 (2003): 219–35.

Des Marais, David J. et al., 'The NASA Astrobiology Roadmap', *Astrobiology* 8 (2008): 715–30.

Giacalone, Joseph A., 'Global Trends in the Commercialization of Space', *The Journal of Business and Economic Research* 6 (2008): 65–76.

Goodpaster, Kenneth, 'On Being Morally Considerable', *The Journal of Philosophy* 22 (1978): 308–25.

Helm, Bennett W., 'Felt Evaluations: A Theory of Pleasure and Pain', *Am Philos Q* 39 (2002): 13–30.

Horowitz, Norman Harold, Jerry S. Hubbard and George L. Hobby, 'The Carbon-Assimilation Experiment: The Viking Mars Lander', *Icarus* 16 (1972): 147–52.

Hubbard, Jerry S., 'The Pyrolytic Release Experiment: Measurement of Carbon Assimilation', *Origins of Life* 7 (1976): 281–92.

Hubbard, Scott, Ken Davidian, Steve Isakowitz, John Logsdon, James R. McMurry, George Nield and Marcia S. Smith, 'Growing the Future of Commercial Space', *New Space* 1 (2013): 3–9.

Jamieson, Dale, 'Animal Liberation is an Environmental Ethic', *Environ Values* 7 (1998): 41–57.

Johnson, Lawrence E., *A Morally Deep World* (Cambridge: Cambridge University Press, 1991).

Johnson, Lawrence E., 'Toward the Moral Considerability of Species and Ecosystems', *Environ Ethics* 14 (1992): 145–57.

Kant, Immanuel, *Groundwork of the Metaphysics of Morals* (Cambridge: Cambridge University Press, 1998).

Kim, Wooseong, Wooseong Kim, Farah K. Tengra, Zachary Young, Jasmine Shong, Nicholas Marchand, Hon Kit Chan, Ravindra C. Pangule, Macarena Parra, Jonathan S. Dordick, Joel L. Plawsky and Cynthia H. Collins, 'Spaceflight Promotes Biofilm Formation by Pseudomonas Aeruginosa', *PLOS ONE* 8 (2013): e62437.

Klein, Harold P., 'The Viking Biological Experiments on Mars', *Icarus* 34 (1978): 666–74.

Kminek, Gerhard and John D. Rummel, 'COSPAR's Planetary Protection Policy', *Space Research Today* 193 (2015): 7–19.

La Duc, Myron T., Wayne Nicholson, Roger Kern and Kasthuri Venkateswaran, 'Microbial Characterization of the Mars Odyssey Spacecraft and Its Encapsulation Facility', *Environmental Microbiology* 5 (2003): 977–85.

La Duc, Myron, Anne Dekas, Shariff Osman, Christine Moiissl, David Newcombe and Kasthuri Venkateswaran, 'Isolation and Characterization of Bacteria Capable of Tolerating the Extreme Conditions of Clean Room Environments', *Applied and Environmental Microbiology* 73 (2007): 2600–11.

Leopold, Aldo, *A Sand County Almanac* (New York: Ballantine Books, 1970).

Levine, Joseph, 'Recent Work on Consciousness', *Am Philos Q* 34 (1997): 379–404.

Lewicki, Chris, Peter Diamandis, Eric Anderson, Chris Voorhees and Frank Mycroft, 'Planetary Resources—The Asteroid Mining Company', *New Space* 1 (2013): 105–08.

Mahnert, Alexander, Parag Vaishampayan, Alexander J. Probst, Anna Auerbach, Christine Moissl-Eichinger, Kasthuri Venkateswaran and Gabriele Berg, 'Cleanroom Maintenance Significantly Reduces Abundance but Not Diversity of Indoor Microbes', *PLOS ONE* 10 (2015): e0134848.

Moissl, Christine, J. C. Bruckner and Kasthuri Venkateswaran, 'Archaeal Diversity Analysis of Spacecraft Assembly Clean Rooms', *ISME* 2 (2008): 115–19.

Newcombe, David A., Andrew C. Schuerger, James N. Benardini, Danielle Dickinson, Roger Tanner and Kasthuri Venkateswaran, 'Survival of Spacecraft-associated Microorganisms Under Simulated Martian UV Irradiation', *Applied and Environmental Microbiology* 71 (2005): 8147–56.

O'Neil, Rick, 'Intrinsic Value, Moral Standing, and Species', *Environmental Ethics* 19 (1997): 45–52.

O'Neill, John, 'The Varieties of Intrinsic Value', *The Monist* 75 (1992): 119–37.

Peeters, Walter, 'Space Commercialization Trends and Consequences for the Workforce', *Acta Astronautica* 53 (2003): 833–40.

Persson, Erik, *What is Wrong with Extinction?* (Lund: Lund University, 2008).

Persson, Erik, 'The Moral Standing of Extraterrestrial Life', *Astrobiology* 12 (2012): 976–84.

Persson, Erik, 'Skapat liv och livets värde', in *LIV – Artificiellt, syntetiskt och utomjordiskt*, ed. E. Persson and J. Abbott (Pufendorfinstitutet, 2017): 219–37.

Plumwood, Val, 'Ethics and Instrumentalism: A Response to Janna Thompson', *Environ Ethics* 13 (1991): 139–49.

Rabinowicz, Wlodek, and Toni Rønnow-Rasmussen, 'A Distinction in Value: Intrinsic and for Its Own Sake', *Proceedings of the Aristotelian Society* 20 (1999): 20–33.

Regan, Tom, *The Case for Animal Rights* (Berkeley: University of California Press, 1983).

Regan, Tom, *Defending Animal Rights* (Urbana, IL: University of Illinois Press, 2001).

Rolston, Holmes, III, 'The Preservation of Natural Value in the Solar System', in *Beyond Spaceship Earth: Environmental Ethics and the Solar System*, ed. E. Hargrove (San Francisco: Sierra Club Books, 1986), 40–182.

Rolston, Holmes, III, 'Duties to Ecosystems', in *Companion to a Sand County Almanac—Interpretive Critical Essays*, ed. J. B. Callicott (Madison, WI: The University of Wisconsin Press, 1987), 246–74.

Rolston, Holmes, III, *Environmental Ethics—Duties to and Values in The Natural World* (Philadelphia: Temple University Press, 1988).

Rolston, Holmes, III, *Conserving Natural Value* (New York: Columbia University Press, 1994).

Rolston, Holmes, III, 'Ethics on the Home Planet', in *An Invitation to Environmental Philosophy*, ed. A. Weston (New York: Oxford University Press, 1999), 107–39.

Sagan, Carl and Sidney Coleman, 'Spacecraft Sterilization Standards and Contamination of Mars', *Astronautics and Aeronautics* (1965): 22–7.

Schon, Samuel C., 'Reversible Exploration Not Worth the Cost', *Science* 323 (2009): 718.

Schuerger, Andrew C. and Wayne L. Nicholson, 'Twenty Species of Hypobarophilic Bacteria Recovered from Diverse Soils Exhibit Growth under Simulated Martian Conditions at 0.7 kPa', *Astrobiology* 16 (2016): 964–76.

Schweitzer, Albert, 'The Ethic of Reverence for Life', in *Animal Rights and Human Obligations*, ed. T. Regan and P. Singer (Englewood Cliffs, NJ: Prentice-Hall, 1976), 133–8.

Singer, Peter, *Practical Ethics* (Cambridge: Cambridge University Press, 1993).

Singer, Peter, *Animal Liberation* (London: Pimlico, 1995).

Smith, Kelly C., 'The Trouble with Intrinsic Value: An Ethical Primer for Astrobiology', in *Exploring the Origin, Extent, and Future of Life*, ed. C. M. Bertka (Cambridge: Cambridge University Press, 2009), 261–80.

Stieglmeier, Michaela, Reinhard Wirth, Gerhard Kminek and Christine Moissl-Eichinger, 'Cultivation of Anaerobic and Facultatively Anaerobic Bacteria from Spacecraft-Associated Clean Rooms', *Applied and Environmental Microbiology* 75 (2009): 3484–91.

Tepfer, David and Sydney Leach, 'Survival and DNA Damage in Plant Seeds Exposed for 558 and 682 Days outside the International Space Station', *Astrobiology* 17 (2017): 205–215

Chapter 10

THE END OF COPERNICAN MEDIOCRITY: HOW MODERN ASTROPHYSICS HAS REINVIGORATED THE SPIRITUAL DIMENSION

Howard A. Smith
Harvard-Smithsonian Center for Astrophysics

Copernican Mediocrity

The Copernican principle of mediocrity was aptly summarized by Goethe in the nineteenth century: 'No sooner was the earth recognized as being round and self-contained than it was obligated to relinquish the colossal privilege of being the center of the universe.' As Carl Sagan put it in the twentieth century: 'We find that we live on an insignificant planet, of a humdrum star, lost in a galaxy, tucked away in some forgotten corner of a universe, in which there are far more galaxies than people' (1980, 4). His claim is that humanity must be an ordinary, commonplace species since our situation is so ordinary. There is the additional implication too, seemingly validated by Darwin, that humanity not only is ordinary, it is also meaningless, simply the evolutionary result of the quasi-random interactions of atoms. Physicist Stephen Hawking expressed the thought this way, his confidence based on the remarkable successes of science from cosmology to genomics: 'We are so insignificant that I can't believe the whole universe exists for our benefit' (Hawking 1995). His belief is not uncommon among scientists and the public, but it often derives from a world view that presupposes such insignificance.

But are we? Modern physics and astronomy have revolutionized the way we conceive of the universe, its size, birth, evolution and content. While its dramatic findings have typically been used to reaffirm Copernican mediocrity, an objective look at just two of the most dramatic recent results – inflationary Big Bang cosmology and exo-planets (planets around other stars) – suggests the opposite. The universe, far from being a collection of accidents, seems (surprisingly) to nurture intelligent life in a framework that is not in the least bit random. This notion, the Anthropic Principle, has been explored for over fifty years and is acknowledged, often grudgingly, by many physicists. The term itself was coined by the physicist Brandon Carter in 1973 and has encountered resistance because of the implication that humanity has something to do with it; 'fine-tuning' is often used instead to discuss the extremely unlikely physical situation in which we find ourselves. Humanity as well, far from being an ordinary lifeform, appears to be extraordinary and could be unique for all intents and purposes given the vast distances between stars and the finite speed of communication or travel. This second result is called the Misanthropic Principle and expresses the observation that in a universe whose physical parameters are spectacularly well suited for life, the environments and circumstances necessary for intelligence to develop are comparatively rare. At least for all practical purposes we are alone. A rational and honest conclusion, therefore, is that humanity is special yet lives in a cosmos that is somehow architecturally suited to produce intelligent beings. We are thus faced with the intellectual challenges of understanding what this could possibly mean and the ethical imperatives that result from being in such a privileged position. For all intents and purposes, as I will review in this chapter, ours is an extraordinary and salubrious cosmic condition, one that is probably cosmically rare and of possible significance. We therefore have an ethical imperative to protect the Earth and its inhabitants.

The first three words of the Hebrew Bible, 'In the beginning God created', have been the source of discussion, debate and doubt for millennia, but arguably the most puzzling word until recently was the first one – 'In the beginning.' Was there even a beginning? Perhaps an eternal Greek cosmos is more perfect and a more accurate description? How did it come to be and what are its cosmological features? Even Einstein thought the universe was unchanging and introduced a 'cosmological constant' into his equations to conform with that picture. Today, however, there are no doubts about this first word: The Big Bang picture, although still replete with incompletely understood knobs (like inflation) and other important puzzles, has been fantastically successful at explaining increasingly precise observations. No, today all the doubt is about the third Hebrew word, God (*Elohim*). Just as science has gradually enabled us to better understand that first word so too, I suggest, does science today offer us new perspectives on the third word, God. One step towards reconsidering that word – and our relationship to the Divine – is encompassed in recognizing and coming to grips with the end of Copernican mediocrity and its two signposts: cosmic fine-tuning and the probable specialness of humanity.

I should conclude this section by adding a point emphasized by the historian of science Dennis R. Danielson: Copernicus and his contemporaries did *not* think his sun-centric system pushed humanity out of a position of glory into one of irrelevance. To the contrary, up until Copernicus, the Greek and Christian views held that the earth was located, as Giovanni Pico put it (c. 1494), 'in the excrementary and filthy parts of the lower world', the very bottom of Aristotle's and Dante's cosmic barrel, where gross, imperfect, mortal beings reside. By putting the Sun at the centre, Copernicus was seen, according to Danielson, as effectively elevating humanity to a place closer to the heavens. During the post-Newtonian era, perhaps partly to appease the religious establishment, philosophers changed their tune: the centre was special and the preferred place to be and Copernicus demoted us to ordinary beings displaced from this 'lofty throne'. The lesson here is that the conventional wisdom, whether about Copernicus's world view, life as a commonplace cosmic accident or Divine purpose, is sometimes in need of reappraisal.

The Misanthropic Principle

The Misanthropic Principle expresses the idea that the multiplicity of possible environments in our cosmos are so varied and uncooperative (or hostile) either always, or at some time during the roughly three to four billion years that intelligent life needs to emerge, that it is extremely unlikely for *intelligent* life to evolve and thrive. It contrasts with the Anthropic Principle, the observation that the physical constants in the cosmos are remarkably finely tuned to make it perfect for hosting intelligent life. Scientists, as a way of guesstimating how abundant intelligent life might be, try to identify all the varied processes needed for an extraterrestrial civilization in the universe and then assign a probability to the chances of each one happening. The most common version of this accounting method is called the Drake Equation, a set of multiplicative factors used to track the various phenomena thought to be necessary to yield intelligent life. However, the Drake Equation is not the mathematical formulation of any physical process.

I review the current science behind the Drake Equation estimates in detailed discussions in my paper, 'Alone in the Universe' in *American Scientist* (2011, 99, 320–7) and more recently in *Zygon* (2016, 51, 497–519). The starting point is the customary one that considers only life capable of conscious, independent thought and an ability to communicate between stars. Primitive life may yet be discovered on Mars; perhaps even multicellular animals will be found on a nearby extrasolar planet. While these revolutionary discoveries would help us reconstruct how life on Earth evolved, unless a species can communicate with us we will be alone – with no one to teach or learn from, no one to save us from ourselves or, in the fanciful extrapolations of sci-fi novels and endless film makers, no one to do battle with. I entitled my paper 'Alone in the Universe' to emphasize this existential solitude. Extraterrestrial intelligence (ETI) implies life able to communicate between stars. We have so far had no confirmed contact with ETI, and as Enrico Fermi famously observed: If they really are common, then where are they?

The most uncertain terms in the Drake Equation estimates for ETI are the three biological ones: the probability that life develops on a suitable planet, that it evolves to be intelligent, and that it survives a long time. All of these involve some element of chance, but the first term seems especially stochastic. In 1970, the biologist Jacques Monod (*Chance and Necessity*) observed that 'Man knows at last that he is alone in the Universe's unfeeling immensity, out of which he emerged only by chance' (180). Stephen Jay Gould, in *Wonderful Life: The Burgess Shale and the Nature of History*, similarly argued that evolution on Earth took a very unlikely path. The Nobel Prize-winning cell biologist Christian de Duve (*Life Evolving*) agrees with them and the role of chance, but argues that chemistry and biology are somehow driven towards making life – but only when the conditions are exactly right. Jonthan Losos, in his new book *Improbable Destinies: Fate, Chance, and the Future of Evolution* (2017), updates Monod's case. Spiegel and Turner (2011) used a Bayesian statistical analysis to estimate the likelihood that life can form from inanimate matter using priors based on the few known facts, and conclude it is highly unlikely. Geneticists have meanwhile discovered that the evolution of DNA was circuitous, and probably the result of many fortuitous accidents. So while it is true that the same physical processes operate everywhere, some sequences of events are astronomically less likely to happen than others. The evolution of intelligent beings could certainly be such a sequence.

An exciting new development in the discussion about ETI comes from the discoveries of exoplanet research, and indeed a key term in the Drake Equation is the frequency of exoplanets capable of nurturing ETI. It has been recognized for a long time, summarized for example by Ward and Browlee in their book, *Rare Earth: Why Complex Life Is Uncommon in the Universe*, that it takes vastly more than liquid water and a pleasant environment to give birth even to simple life, much less to complex life or ETI. At a minimum it also takes an environment stable for billions of years plus all the right ingredients (the abundances of elements, for example, are not uniform across the universe). Since the time of the Greeks, we have expected planets to be common. Until just a few decades ago most scientists and textbooks agreed that practically all moderate mass stars hosted exoplanetary systems like our solar system, with an earth-like planet capable of bearing intelligent species. As Goldsmith and Owen put it in their 1993 textbook, 'Nothing in our theories for the origin and evolution of our sun is unique to the solar system. … The chances seem good that one of these inner planets will orbit its star at the right distance [to host life]. … We say one in every two to be conservative' (Goldsmith and Owen 1993, 384). The news is that the latest results of astronomy show otherwise. So far over 4200 exoplanets have been detected; many of them have had their masses, radii and other properties measured. The single most remarkable discovery about exoplanets is their exotic variety. Many are located in highly elliptical orbits around unstable stars, making evolution over billions of years difficult if not impossible. Other systems contain giant planets that may have drifted inward, disrupting other planets; the majority of exoplanets seem to be around low mass stars whose X-ray emission and stellar winds are usually inhospitable, and there are many other, unanticipated properties.

The discovery of earth-*sized* exoplanets is a remarkable technical achievement worth special comment since so many people assume earth-sized means earth-*like*. It doesn't by any stretch. So far about a dozen earth-sized exoplanets have been found and many more will added to the list once the NASA's TESS satellite is launched in 2018. Some of these are in environments (temperatures, stability, etc.) making them suitable for life. At the start of 2018, there were eight candidates whose sizes are less than about two earth-radii, that have rocky cores (thought to be a key prerequisite for life), and that are also located at a distance from their star at which any surface water could be liquid (the so-called habitable zone, although no surface water has yet been detected). But even on most of these select bodies the emergence of life (not to mention its evolution to intelligence) faces severe difficulties: a few are bathed in extreme X-rays from their stars (a common problem), and none have evidence of an oxygenated atmosphere. This is not to mention many of the other issues that impede life on planets, earth-sized or otherwise, even life that might take on weird, non-carbon-based forms (see Smith 2011; 2016). The bottom line for ETI appears to be that they are probably much rarer than previously imagined. There are, on top of this, some unbreakable physical realities to confront if we want to find them, in particular the finite speed of light.

Still, some aliens might be nearby. The Sun lies in a cavity of interstellar gas called the Local Bubble that extends over roughly 600 light years. It in turn is located in Gould's Belt, a spur of stars, star clusters and molecular clouds between two of the Milky Way galaxy's spiral arms, stretching from the Orion nebula to the Ophiucus-Scorpius clouds and on to the Perseus clusters – a distance of about 1,250 light years in its longest dimension. It takes a light signal 1250 years to traverse it; our earliest broadcasts have only made it out to about 100 light years. What might be the chances of chatting with alien neighbours if we wait and listen for the next 2500 years, long enough for our message to reach the farthest neighbourhood and their reply to return to us? If a human generation is twenty-five years, then in over one hundred generations we could converse with all aliens throughout this neighbourhood. The approximate number of stars per cubic light year in this volume here is 0.004, to within a factor of two, or about 30 million stars of all types in a volume of radius 1,250 light years. The combined chances for ETI existing, therefore, need to be better than about one in thirty million. If the stars are too big or too small, if the planets' orbits or obliquities are wrong, their sizes or chemical compositions unsuited, their surfaces ill-equipped, their geologic and meteoritic history too inauspicious, then we are alone. Then add in the biological uncertainties, which are much less well understood: If the chemistry needed to generate life is too intricate or too slow, if evolution from proteins to intelligence is too often aborted or misdirected, or if civilizations die off quickly, then, too, we are alone. If we choose to examine a volume one hundred times smaller than that enclosed within a 1250 light years radius, that is accessible within a single generation, we will have a yes-or-no answer much sooner, but the chances of success go down by a factor of a million because the number of stars is proportional to the volume of space and scales with time (distance) cube. If we expand the search volume to improve the chances of finding other beings, the wait time goes up correspondingly. For a very long time, then, it seems likely that we will be alone. It does not matter whether or not aliens thrive in the distant reaches of space – they might or they might not. What matters is whether we can communicate them – until then we are just guessing.

The SETI programme (Search for Extraterrestrial Intelligence), which undertakes extensive searches for signals, fully acknowledges this science but argues that 'the only significant test of the existence of extraterrestrial intelligence is an experimental one' (Tarter 1983, 359). I support SETI in principle – if we don't look for them we won't find them. But it is a risky endeavour, and non-professionals especially need to be reminded of the enormous limitations imposed by the finite speed of light. Even if there were a network of advanced civilizations living on the other side of our galaxy, they much are too far away from Earth to converse with, and to be 'alone' is to be without anyone to talk to. There is one important new feature of SETI to mention. When it started over fifty years ago, SETI searches used what today we would consider relatively basic technology and probably could not even have spotted the Earth from a distance of 100 human generations of light-travel time. Today, technology is not the limit. SETI both has very much better sensitivity and new sources of funding like the Breakthrough Initiative that hopefully will provide steady support. According to the programme's goals, the Initiative could detect a civilization around the 1,000 nearest stars if it transmits towards us even with only the power of common aircraft radar, and it could detect a civilization transmitting from the centre of the Milky Way (more than twenty-five thousand light years distant) if it broadcast with more than twelve times the output of our current interplanetary radars. We have seen nothing yet, but in much sooner than a thousand-year wait time even null results should enable us to reach statistically significant conclusions about any ETI that are transmitting signals.

Even if the formation of life were inevitable on every planet in the universe with liquid water, and even if the Milky Way galaxy has millions of water-bearing earth-sized planets, *for all practical purposes* we and our descendants for at least 100 generations are living in solitude. We are most probably alone. To recognize this state is to have a renewed appreciation for our good fortune and to acknowledge that life on Earth is precious and deserves supreme respect. Humanity is not mediocre.

The Anthropic Principle

The Anthropic Principle has been contemplated for decades, since theoretical physicist Paul Dirac first called attention to the curious balance between large cosmic numbers. We live on a planet with liquid water, located at just right distance from the sun so that the surface temperature enables water from being completely frozen or totally evaporated – the so-called habitable zone of the solar system. The Earth is hospitable for many other reasons as well: It has a conducive chemical makeup, tectonic plates to rejuvenate its oxygen-rich atmosphere, a tilted axis for seasons, a large moon to stabilize its tilt, and many more wonderful properties. No place else in our solar system is even close to being like it or capable of hosting intelligent life. We seem to be fabulously lucky. But the universe is a big place, with trillions and trillions of stars. Probably a paradise like Earth will randomly happen somewhere. In this way of thinking, we are not lucky at all – we are here because this is where we can be. It's about randomness and having lots of choices.

Now consider the universe. Its fundamental constants take particular values – the speed of light or the strength of the nuclear forces, for example – but why these values? We have no idea. But we do know that if those numbers – which in principle could take any values large or small – were much different from what they are, we would not be here. If, for example, the electromagnetic strength were only a few per cent different, then water would not be a good solvent nor an essential ingredient of life. If the nuclear force were tens of per cent weaker then atoms of carbon – critical for constructing complex molecules and life – would be much rarer. The most extreme example of fine-tuning is the expanding universe itself. Physicists estimate that if the balance between cosmic effects were different by only 1 part in 10-to-the-power 120 (!) we would not be here. Barrow and Tipler's exhaustive 1986 book, *The Anthropic Principle*, is one of many detailing these amazing 'coincidences' about the physical structure of the universe and its miraculous suitability for life. Geraint Lewis and Luke Barnes, in *A Fortunate Universe: Life in a Finely Tuned Cosmos*, re-examine Barrow and Tipler's arguments and offer some new ones. Meanwhile, the idea remains controversial in many circles, and there is disagreement about exactly how fine-tuned the constants really are. Fred Adams of the University of Michigan, for example, has recently calculated some much less restrictive scenarios. But at least for now there is strong evidence for fine-tuning, raising the question: Why? Why is the universe so perfect?

There are so far only three answers from science. One is just dumb luck. The second answer, proffered and defended by some theoretical physicists, is that there are an infinite number of universes – a 'multiverse' – spanning all logical possibilities. We just live in the one we can; no big deal. The third answer touches on philosophy, and comes from quantum mechanics. Matter is composed of wave functions of probability that only become 'real entities' when they are measured by a conscious observer. The quantum mechanical pioneer John Wheeler famously proposed that the universe *had* to evolve conscious beings in order to become real, and this notion (sometimes called the Participatory Anthropic Principle) is very much alive in modern texts. Moreover, consciousness and its origins are mysteries in their own right. Complex systems, for example, can produce unexpected phenomena through what are called 'emergent' processes. Consciousness, some physicists propose, is just such an emergent phenomenon. Emergent phenomena are real, but emergence still remains an after-the-fact justification rather than an explanation for how consciousness arose, with more work needed to make it convincing. The arguments by philosopher and theologian John Haught (*Is Nature Enough?*) provide one cogent view of the explanatory limits of emergence.

I admit to not being a fan of any of these three. It seems a cop-out to say we are just lucky, and as a physicist trained to give preference to simple solutions, a multiverse strikes me as way too extreme. The quantum mechanical route is possible, but uncomfortably mysterious – yet there are quite a few mysteries still in quantum mechanics so of the three it has potential. The point is that if some process – perhaps quantum mechanics but maybe something else – steers the universe toward producing intelligence, then we humans are representatives of that endpoint. It suggests that we play some cosmic role. Finally, if we might be the *only* such intelligent beings around (or that we will know about for millennia or longer) then we not only are not mediocre, but are cosmically special. Many people will reject this notion. The remarkable successes of science and the fact that the same laws of physics apply throughout the universe have persuaded them that whatever processes that produced intelligence on Earth must be operating everywhere, and therefore life must be everywhere. I certainly agree with the first part, but as noted earlier some outcomes can be astronomically less likely than others.

The End of Copernican Mediocrity

People with an unwavering faith in the existence of ubiquitous extraterrestrial civilizations and the conviction that humanity cannot be special might assert that we are mediocre, but beliefs and convictions are not proofs. There are many examples of famous scientists making claims founded on personal beliefs that we now see as ridiculous. Percival Lowell, for example, known for his search for Pluto and studies of the canals of Mars from his Flagstaff observatory, wrote in his 1908 book, *Mars as the Abode of Life*: 'From all we have learned of its constitution on the one hand or of its distribution on the other we know life to be as inevitable a phase of planetary evolution as is quartz or feldspar or nitrogenous soil. Each and all of them are only manifestations of chemical affinity' (1908, 37). No one today thinks this. Every schoolchild knows that Mars has no artificial canals and no aliens either. Lowell was by no means unique. Before him, in 1803, the eminent French astronomer Jerome Lalande had confidently written, 'Is it rational to suppose the existence of living and thinking beings is confined to the earth? From what is such a privilege derived but from the groveling minds of persons who can never rise above the objects of their immediate sensations?' (Lalande, quoted in Fontenelle 1803, viii). More recently, Harlow Shapley (1885–1972), the distinguished director of my own institution, the Harvard College Observatory, wrote of 'intimations of man's inconsequentiality' in a vast cosmos and of 'our [firm] belief in the cosmos-wide occurrence of life' (Shapley 1963, 3, 77). Lowell's confident assumption and Lalande's rhetorical logic were unproven, and their arrogant assertions were just wishful thinking. Michael Crowe has reviewed how the assumption of the existence of cosmic aliens, perhaps originating with the Greeks, was pervasive by the seventeenth century among both scientists and theologians (Crowe 1997; 2008).

The simplest and most rational explanation for both the Anthropic and Misanthropic Principles, consistent with the observations so far, is that humanity is not mediocre. Indeed, the evidence to date suggests we could be exceptional – at least as far as we will know for a very long time and, since we live in an ageing universe in which galaxies are moving apart at an accelerating rate, perhaps forever.

Three Dilemmas

The Misanthropic Principle raises three acute dilemmas that have not yet been carefully explored: epistemological, theological and ethical (Smith 2016). The epistemological dilemma is clear: Not knowing about the existence of something does not mean it does not exist. Until we hear a clear signal from beyond, or until our science has progressed far enough to provide some kind of all-embracing and conclusive answer (although the nature of such absolute evidence is hard to imagine), humanity is left in an existential quandary. In this environment of necessary ignorance, how should scientists and theologians respond to the many people for whom the prospect of being 'alone', without hope for salvation or comfort from a super-intelligent species abiding in heaven, is frightening? Having been the Chair of Astronomy at the National Air and Space Museum for ten years and having interacted with tens of thousands of visitors, I would stress to my scientific colleagues the importance of not overstating (or worse – misrepresenting) the significance of exoplanet discoveries. The temptation is great to inject a few tease words about 'life' into every report of a new planet found in its habitable zone. Yes, it makes the boring technical details seem more exciting. But it is vitally important not to 'cry wolf' too often lest the astronomy community and science as a discipline lose credibility. The excitement of exploring and analysing strange new worlds is more than justification enough.

I might add that we moderns now understand that unknowability is fundamentally allowed by the character of our universe. Chaotic systems, for example, even relatively simple ones like our solar system, do not allow practical predictions of outcomes arbitrarily far into the future. Although the physical laws are well known, the equations explicit, and cause and effect rules are clearly understood, the outcomes are unknowable. Other fundamental unknowables are events occurring beyond the cosmic 'horizon', the farthest light allows us to see in the universe in its lifetime (13.8 billion years). This applies even if these events have already happened. Moreover, in our outwardly accelerating universe more and more space is crossing over that dark horizon. Most non-intuitive of all – although quantum mechanics allows us to calculate precisely how a wavefunction will evolve, it does not allow us to know how it will eventually materialize. Not being able to know whether or not we are alone in the universe has a similar flavour, however it not only touches us on a personal level, it teases us with the possibility that we could someday find a positive answer.

The second problem is theological. For the community of spiritual believers, the conclusion that we could be special might on first glance appear to be reassuring. For at least the past century, however, theologians of many religions, and especially Christianity, have worked hard to include the idea of extraterrestrial life into normative religious thought. Their consensus opinion can be summarized by Bishop Krister Stendahl who phrased it, 'It seems always great to me, when God's world gets a little bigger [that] I get a somewhat more true view of my place and my smallness in that universe.' God's power is glorified, not diminished, by an abundance of life. If, however, humanity is singular, the theological community must adopt a much different perspective on Divine potency and the world, not to mention on the significance of humanity and the implications for human fulfilment. I very much admire the humility of Bishop Stendahl, but it is also humbling to think that we are, perhaps, special beings. Moreover, if we might be unique (at least as far we will probably know for millennia) then we must reconsider the possibility that we are not an accident but were created by some kind of intent, even for some purpose.

For atheists and scientific reductionists, the theological dilemma seems particularly grave (though most people will share some of the angst). We moderns are nearly convinced by the Epicurean argument, a feature of reductionism, that we are a randomly evolved collection of atoms. If we are not – then what? For atheists confronting this disorienting challenge, modern physics and philosophy have offered at least two solutions. The Quantum Mechanical solution mentioned above includes the still incompletely understood implication that the world and its matter are composed of wave functions of probability that only become real entities upon being measured. The so-called Copenhagen Interpretation has consciousness beings as the source of the measurement. The most current alternative quantum mechanical interpretation invokes the principle of decoherence, in which the quantum state is quickly fixed by a multitude of environmental interactions, but other physicists respond that this does not resolve the philosophical issue. Meanwhile modern philosophers of consciousness and the mind like Thomas Nagel have argued, similar to Wheeler, that a still-mysterious, but apparently essential, aspect of nature results in the development of conscious life. The natural processes for generating life may be mysterious but might still produce it frequently. Or it might not. For either group, atheists or theists, true believers in human ordinariness might just decide to hold fast to their current dogmas and adopt a wait-and-see attitude for a few more millennia.

The third dilemma is the ethical one, and I argue that it cannot wait. The Earth itself is under stress, and humanity is confronting growing misery. Shall we turn away while species that have taken 13.8 billion years to develop become extinct, or while our changing climate radically alters the environment for life? If we are merely a collection of evolved atoms, then these issues are of no great concern. The argument would be: There probably is life elsewhere distributed among the stars, abundant and diverse, along with many salubrious, earth-like planets to which we might escape but that anyway may have civilizations of their own of equal value to ours; some of these alien civilizations will survive even if the Earth's doesn't – perhaps that is enough. But what if they are wrong? If the human race – as far as we are likely to know for millennia – is alone, we must face the possibility that the above is not true, and neither we nor our planet are products of common happenstance. The Earth and its life have value. This prospect brings great urgency to the cause of protecting our rare planet and all of its precious inhabitants. We humans are at least unusual, and we are certainly blessed. Since biblical times, blessings carry with them added responsibilities and concomitant consequences. Those responsibilities include the obligation to deal compassionately with other beings and to attend to the welfare of community and its environment.

The Jewish view of the state of being blessed offers some insights into these three dilemmas. The epistemological issue has no practical implications and is not particularly problematic: Whether or not we know about others, we know about our own blessings. Indeed, says the prophet Amos, urging humility: 'Are ye not as children of the Ethiopians unto me, O children of Israel? saith the Lord. Have not I brought up Israel out of the land of Egypt? And the Philistines from Caphtor, and the Syrians from Kir?' (Amos 9.7). There might be other intelligent beings in the cosmos, but our obligations are independent of theirs. The theological issues are similarly unproblematic. The absence of ETI in our sphere of influence only enhances our self-awareness of our peculiar status, and their possible presence does not diminish it. The ethical dilemma is the one for which I think a Jewish perspective is the most helpful. Our exceptional status on Earth, and our new-found awareness of this probable good fortune, should make us more sensitive of our task 'to serve the Earth and to protect it' (Gen. 2.15). 'When you live in the land that flows with milk and honey … you shall therefore obey the voice of God and keep his commands' (Deut. 27.3). The Jewish perspective not only asserts that we should try, it emphasizes that we have the skills to succeed. But is the herculean task of caring for humanity, life and our fragile planet beyond our abilities? Could an atheistic inclination lead to pessimism? Things are what they are, and whether or not humanity perishes makes no particular difference to the cosmos. Why bother? The first-century Rabbi Tarfon offers a famous aphorism that provides a basis for positive motivation rather than despair. It is grounded in the religious notion that we are blessed and therefore obligated to assume responsibility: 'You are not expected to complete the task', he writes, 'but neither are you free to abstain from the effort' (Talmud Avot: 2, 15). And, if perchance we are not alone, then we share in the cosmic goals of making the world better, *tikkun olam,* with all other conscious beings with free will, although we may never know about them for sure.

Being Human

> We find ourselves in a bewildering world. We want to make sense of what we see around us and to ask: What is the nature of the universe? What is our place in it and where did it and we come from? Why is it the way it is?
>
> – Stephen Hawking, *A Brief History of Time*

> Where were you when I secreted matter? Speak up if you understand … how the flow was contained or how it burst forth!
>
> – Job 38.4 (translation from Smith 2006)

Today we, unlike Job, can tentatively raise our hands to God. With humility we can respond to this rhetorical question, 'Well, yes, we think we might have an understanding of these things.' Unlike Job, today we can believe in a God *not* because we are ignorant, but because we understand. The 'god of the gaps' is the derogatory term for the deity invoked to explain those features of the world left incomprehensible by gaps in theories. When creation ex nihilo was a mystery to science, God was needed to provide an explanation; likewise God was invoked to explain the near perfection of living creatures when that was a mystery. Stephen Hawking advocated an interesting and more physical variant to this notion – the god of boundary conditions, what he called the 'god of the edges'. He noted that modern physical theories, which strive for completeness, have been excellent at explaining (more or less) all aspects of a complete system like the universe as it is evolving. What they cannot easily do is explain the boundary conditions. 'If there is an edge', he explains, 'somebody has to decide what should happen at the edge. You would really have to invoke God.' (Hawking introduced the idea in the general context of his theory that the universe has no boundary/edge in time, even at the moment of creation.)

I have written about Jewish mystical approaches to cosmology and the creation in *Let There Be Light: Modern Cosmology and Kabbalah, a New Conversation between Science and Religion* (Smith 2006). The Kabbalah is not an alternative to Big Bang cosmology, although it wonders about many of the same things, but it is a source of new language and a different way to think about related cosmic things, consciousness, for example, balance and ethics. For the Kabbalists, Hawking's variant with its relegation of God to the edges has it backwards. It is the intimate, unbounded *wholeness* of the world that is the salient attribute that signifies God, not its bounded finitude.

Once people realized that the cosmos was not geocentric, they began to think about themselves, humanity and their world in a new way. The modern evidence for the end of Copernican Mediocrity, the Misanthropic Principle, should initiate a similar process of self-reappraisal. We seem to be unusual and possibly even unique, although we are unlikely to know for sure one way or the other for a very long time, perhaps forever. Still, it is possible that we are just an accident, with no particular significance. But conscious life appears to be a remarkable and unanticipated achievement of the universe – not an attribute one would have predicted for an ensemble of atoms. Even if we are not unique (though we may never know for sure one way or the other) we should admit that the bias underlying the modern preference for mediocrity – namely, that we are nothing more than a random accident – may no longer be viable. The Anthropic Principle intimates that some necessary feature of nature endowed the cosmos with this capacity, making it fundamental to the Big Bang creation and steering it over aeons of evolution to produce conscious beings today. If so, then we are representatives of that teleological endpoint, and serve a cosmic purpose of extraordinary significance. The philosopher Thomas Nagel puts it this way in his 2012 book, *Mind and Cosmos*: 'We have not observed life anywhere but on earth, but no natural fact is cosmologically more significant' (32).

The arguments for or against an end to our cosmic mediocrity necessarily rest on statistics, incomplete data and the admission that there are many things that we still do not understand. Research will continue to make progress in quantum physics, and in the search for basic life on exoplanets too, but the evidence that humanity is precious is likely to remain compelling. Should we not, therefore, treat one another as the priceless beings we seem to be? The Earth, even if turns out not to be unique, is for all intents and purposes a special place – should we not care for it as we would the most precious of our family heirlooms? The implication of the Anthropic Principle is that it matters. The implication of the Misanthropic Principle is that we will have to assume these awesome responsibilities by ourselves, without help from alien insights or technologies. Modern science may have prompted this re-evaluation, but addressing it will require the best of all our human abilities.

We live in an extraordinary time in which seemingly every mystery has become or is becoming intelligible. The 'god of the gaps' – the god of mystery – is no longer the default explanation for even that archetypical riddle: the Creation. But science in our miraculous era, although it refashions the inscrutable into the comprehensible, simultaneously transmutes the mundane into the wonderful. God is a Deity of wonder as much as a Deity of mystery; such is the import of Scientific Understanding, the fabulous – the real – Tree of Knowledge.

References

Adams, Fred C., and G. Evan, 'On the Habitability of Universes without Stable Deuterium', *Astroparticle Physics* 91 (2017): 90–104.

Barrow, John D., and Frank J. Tipler, *The Anthropic Cosmological Principle* (Oxford: Oxford University Press, 1988).

Crowe, Michael J., 'A History of the Extraterrestrial Life Debate', *Zygon: Journal of Religion and Science* 32 (1997): 147–62.

Crowe, Michael J., *The Extraterrestrial Life Debate: Antiquity to 1915* (Notre Dame, IN: University of Notre Dame Press, 2008).

Goldsmith, Don, and Tobias Owen, *The Search for Life in the Universe*, 2nd edn (Reading, MA: Addison-Wesley, 1992).

Haught, John F., *Is Nature Enough?: Meaning and Truth in the Age of Science* (Cambridge: Cambridge University Press, 2006).

Hawking, Stephen, *A Brief History of Time* (London: Bantam Books, 1988).

Hawking, Stephen 'Interview with Ken Campbell', *Reality on the Rocks* (1995).

Laland, Jerome, *Conversations on the Plurality of Worlds*', ed. B. Fontenelle (London: J. Cundee, 1803).

Lewis, Geraint F., and Luke A. Barnes, *A Fortunate Universe: Life in a Finely Tuned Cosmos* (Cambridge: Cambridge University Press, 2016).

Losos, Jonathan B., *Improbable Destinies: Fate, Chance, and the Future of Evolution* (New York: Riverhead Books, 2017).

Lowell, Percival, *Mars as the Abode of Life* (New York: Macmillan, 1908).

Sagan, Carl, *Cosmos* (New York: Random House, 1980).

Shapley, Harlow, *The View from a Distant Star* (New York: Basic Books, 1963).

Smith, Howard A., *Let There Be Light: Modern Cosmology and Kabbalah, A New Conversation Between Science and Religion* (Novato, CA: New World Library, 2006).

Smith, Howard A., 'Alone in the Universe', *American Scientist* 99 (2011): 320–7.

Smith, Howard A., 'Alone in the Universe', *Zygon* 51 (2016): 497–519.

Spiegel, David, and Edwin Turner, 'Life Might be Rare Despite Its Early Emergence on Earth: A Bayesian Analysis of the Probability of Abiogenesis', *Publications of the National Academy of Science* 109 (2012): 395–400.

Tarter, Jill, 'SETI Program', *Science* 220 (1983): 359.

Ward, Peter D., and Donald Brownlee, *Rare Earth: Why Complex Life Is Uncommon in the Universe* (New York: Springer-Verlag, 2000).

AFTERWORD: OUR PLACE IN THE UNIVERSE

Tom McLeish
York University

The Science Museum in London hosted a remarkable exhibition in 2016: *Cosmonaut* celebrated the Soviet Russian space programme from its inception to its abortive moon-landing project, exhibiting many artefacts never seen in public before in any country, let alone in Western Europe. As a scientist inspired in childhood by the 'space race' of the 1960s, I confess to wandering around the galleries gazing rapturously at the Soyuz craft, space suits, rocket parts – and finally at the prototype Russian Lunar Module from which, in an alternative history, Alexei Leonov would have been the first human being to step out onto the lunar regolith. But the first chamber of the exhibition was thought-provoking in a different way. Dedicated to the philosophical background that drove the Russian space programme from the first rocket trials of Tsiolkovsky, it featured the nineteenth-century *Cosmic* movement of N. Fedorov and others (Young 2012). For it was their romantic-theological vision, rather than any international competition, that triggered the technological reach into space for the Russians. The cosmists drew on deep Orthodox Christian roots: for them the doctrine of universal resurrection implied the breaking of spatio-temporal bonds. The eventual antithesis of finite mortality in time and the confines of Earth in space becomes the human habitation of the whole cosmos. I always found it remarkable that the ostensibly atheist Soviet Union named their early spacecraft *Vostok* (east) and *Voskhod* (dawn or ascent), with unmistakable resonances of resurrection, while the combative American programme chose pagan deities (*Mercury, Apollo*).

It might be objected that even if the past origins of human spaceflight have theological resonance the thinking within the chapters of this volume is directed towards the future, not the past. Thinking about the future has its own history, however, and the cosmists were also the earliest of the technological futurists, anticipating artificial cloning, the trans-human and artificial immortality. Their ethical concerns, or their critics, were principally focused on the consequences of unfettered occupation of the cosmos on the human, rather than on the non-human cosmos, as this collection. Paradoxically, in an anthropocentric concern for the consequences that our actions have for rest of the world, and possibly for other life and intelligences, we frequently fail to notice the huge consequences of a cosmic future for ourselves and for the choices we will have to make in its light. Fabian Revol comes close when he points out the necessity of 'critiquing the principle of plenitude'. Only a vanishing fraction of the hyper-astronomically large universe of possibilities can be realized. Our human choice is to constrain the plenitude of imagination, desire and potency with the form of wise choices. The time for such choices is approaching faster than some might think – just recently Elon Musk's commercial launch company SpaceX propelled, in a bizarre yet portentous demonstration of astro-kitsch, his own car into a deep-space orbit that would intersect closely with Mars. (At stake here was the integrity of the 'biosphere' of Mars, which Erik Persson explores in his chapter.) Some brave sallies into theological thinking around potential future engagement with other cosmic intelligences have already been advanced, such as the recent work by David Wilkinson (2013).

There are three big contextual questions that have been suspended in the background of the previous chapters. They are essentially relational, delineating the mutual standing of humankind and the cosmos as a whole, and addressed in different ways by this volume's authors. First is the search to resolve whether our future place and role in the cosmos matters to anyone or anything beyond ourselves. (I take it that all agree it is significant to us.) This is the question of our significance, raised and discussed in fresh ways by Howard Smith. Second, as Smith also implies, we need to anticipate whether, on a cosmic scale, the human race is potent or impotent to change nature significantly beyond the limits of Earth. This is the question of our capacity. Third we need to establish the moral dimension to this engagement, if there is one. If, in other words, through potency and significance we are in a place to alter or to inhabit the universe beyond Earth, what are the moral frameworks available to us to construct in the Kantian dialectic between our evolving inner and outer worlds? A fourth question perhaps peers more tentatively from the undergrowth – the one that asks whether the balance of action along the axis from the human to the cosmos is predominantly active or passive. It still has not taken centre stage at the cosmic level because it is only late in the last century that one could be taken seriously when asking whether more damage could be wreaked by the planet on humankind or vice versa. We are just grappling now in the first decades of eco-philosophies, theologies and politics that recognize the reciprocity of the relationship of humans and nature within our world of origin. So Erik Persson is right to ask, from this point of view, whether the ethics of extra terrestrial life can operate as an extension of a terrestrial ecology, or calls for categories and choices that are entirely new.

The case for the science-fiction novel as an arena for exploring such choices and anticipating their consequences and dilemmas is persuasively put by Zoë Lehmann, who takes the very 'terraforming' of Mars advocated by Musk as her topic. She shows that her authors have thought deeply into Martian eco-technological future, and brought characters to life who embody the arguments for and against deliberate anthropogenic transformation of the natural world. 'Hands off sacred nature' and 'Apply the tech-fix' tug at each other over the greening of the Red Planet, but are of course also two poles of a spectrum of narratives that weave their way through the public and political context for science in our own time, and right here on Earth, as well. Philosopher J.-P. De Puy has commented on the ubiquity of five 'narratives of despair' (Dupuy 2010) that propel public discussion of scientific topics, and the ethical questions they give rise to. They find localized form in different contexts; a survey of European reactions to nanotechnology, for example, expressed them in this way (Davies, Macnaghten and Kearnes 2009; see also McLeish 2015):

1. 'Be careful what you wish for' – the narrative of Desire
2. 'Pandora's Box' – the narrative of Evil and Hope
3. 'Messing with Nature' – the narrative of the Sacred
4. 'Kept in the Dark' – the narrative of Alienation
5. 'The rich get richer and the poor get poorer' – the narrative of Exploitation.

These are the suspicions with which the 'technological fix' are held, but often swim underneath the surface of public debates, just as buried themes of longing and loss swim beneath the surface of a novel. We will return to the narratives of despair later, but for now note that they have their own roots in ancient or modern pagan or atheistic outlook; none even glimmers with hope, resurrection or of Revol's echoing Aquinas' trinitarian vestiges. Those who might have been weaving a theological narrative for science and technology seem to have been strangely quiet, at least as the narratives of public media are concerned.

It is fascinating to review the emergence of the novel (and we should also say since the second half of the twentieth century, the film) as a vital tool in anticipating and instantiating the ethical debates that a cosmic future presents. Perhaps only in robotics, thanks to Asimov, has this creative genre been taken seriously as a research tool in 'futures' thinking. It would not have surprised the nineteenth-century French novelist Émile Zola, however, whose 1880 essay *Le Roman Éxperimentale* responds to his agitated and entranced reading of a contemporary biomedical text (Zola 1964). Claude Bernard's earlier *Introduction à l'étude de la médecine expérimentale* urged that the methodology of the natural sciences be applied to the entire human being. Zola extrapolated in animated fashion, seeing a hierarchy of emergent scales in human nature that would fall, one by one, to the march of science. If a molecule, why not an ensemble of molecules; if a human individual, why should not an entire society be amenable to the scientific method? These are the questions he poses, as he sets out an imaginative project that we would now (almost) recognize as the social science of Weber, Marx and Deleuze. Here he explains the heart of his project:

> And this is what constitutes the experimental novel: to possess a knowledge of the mechanism of the phenomena inherent in man, to show the machinery of his intellectual and sensory manifestations, under the influences of heredity and environment, such as physiology shall give them to us, and then finally to exhibit man living in social conditions produced by himself, which he modifies daily, and in the heart of which he himself experiences a continual transformation. Thus, then, we lean on physiology; we take man from the hands of the physiologist solely, in order to continue the solution of the problem, and so solve scientifically the question how men behave when they are in society.

His idea of the experimental novel is one based on observational reality, but it is also a novel, a place of invention and creation. We do not now take it as coincidence that the early novel and experimental science arose within a generation (of Boyle and Defoe) in the seventeenth century. Both forms are imaginative, artificial constructs that isolate a set of actors and constraints, extracted and idealized from nature, and enquire of the outcome. There is a common and original conception in that critical early modern period that the artificial and confined need not be an irrelevance to the natural and entire. We can indeed learn about nature from its reduction to small subsets; the in vitro teaches us about the in vivo. The difficulty that faces us now is to understand how very counter-intuitive this was in earlier centuries when the running metaphor for nature was a book to be *read* not *written*. When Galileo referred to 'God's two Books' he was not at that point innovating, but invoking a trope from the twelveth-century philosopher and theologian Hugh of St Victor (Hugh of St Victor *De tribus diebus* 4, PL 176, 814B). Writing nature now is what experimental science does; writing possible futures is to experiment with them before they are realized.

To 'write nature' takes imagination. And imagination in turn, as the medievalists also knew better than we do, draws on aesthetics as a source of, not just as a response to, art. So Kurt Willy Saether is right to draw our attention to the role of beauty, the sublime and wonder as active agents, or 'seeds of science' as Emerson put it. Perhaps the relative silence of science on the intrinsic function of the aesthetic today is one of the reasons that its public narratives are so grey, so dismal. A long view of the history of natural philosophy and its metaphysics is as necessary to a well-supported preparation for our scientific future in the cosmos as the deep roots of a lofty tree are to the stability of its upper storey. Among the many disciplinary divorces to be lamented over, that between science and its own history is perhaps the most tragic.

In the early thirteenth century, the English polymath Robert Grosseteste set out to describe the purpose of learning itself in the medieval curriculum of the seven liberal arts. In his preface he draws on earlier reflection by Anslem of Canterbury on the complementary roles of cognition and emotion:

> The works within our capacity consist either in the mind's sight, or in the desire of the same, or in bodily motions, or in the dispositions of these same motions. Sight first looks; then it verifies what has been looked at or cognized, and when the fitting or harmful have been verified within the mind or within sight, desire strains to embrace the fitting, or retreats into itself to shun the harmful.

There has been very little reflection on the dense interplay of emotion and cognition at this depth of self-contemplation since the century that Grossteste ushered in, and Roger Bacon out. Only very recently, under the auspices of neuroscience, has there arisen a similar notion of cognitive duality, conjecturing two distinct processes of, 'System 1' and 'System 2' (Stanovich and West 1999; Evans and Stanovich 2013). The first is fast, autonomous, intuitive and integrative; the second serial, conscious and rule-based. The only other modern expression of the deep role of affect in the conception of analytical thinking is due to the mathematician Henri Poincaré, in a celebrated and extended meditation on how mathematical creativity works (Poincaré 1915). Poincaré gives an extended personal account of the familiar pattern of hard conscious struggle and apparently fruitless labour, before setting the problem down and, again apparently, leaving it, only to experience at an unforced moment the conscious appearance of a beautiful new approach towards a solution. But then he goes further than others who have told similar stories – charging himself to think seriously about how the non-conscious mind achieves such a miracle.

> What is the cause then, among the thousand products of our unconscious activity, some are called to pass the threshold, while others remain below? Is it a simple chance which confers this privilege? Evidently not; among all the stimuli of our senses, for example, only the most intense fix our attention, unless it has been drawn to them by other causes. More generally the privileged unconscious phenomena, those susceptible of becoming conscious, are those which, directly or indirectly, affect most profoundly our emotional sensibility.

This is a remarkable idea, implying that it might be meaningful to ascribe dual categories such as logical thought and emotion, or cogitation and affect, to non-conscious as well as conscious mental processes.

Our future place in the cosmos will draw on our future discoveries about the structure and capacities of the human mind – an adventure that will surely need to explore the extraordinary creative potential of the sub conscious, where reason and emotion know no separated meaning. The future is not only about where we go, it is about how we think, and about what we know about how we think.

The 'theory of mind' is the celebrated, and disputed, prime philosophical territory of 'emergence', the pivotal point of Conor Cunningham's contribution, which urges an antithesis of a reawakened 'scientia' against neo-reductionism. His elegant articulation of the 'macro' enslaving the 'micro', in robust opposition to the habitual assumption of supervenience, needs repeating and developing. But here we find ourselves again ill-equipped to grasp the receding horizons of our future potential by the disciplinary divisions of our present academy. It is essential that we understand the nature of the cosmos that we contemplate populating. For the question of whether we inhabit a reducible or emergent universe is surely as much a question of science as of philosophy. One could imagine, even construct mathematical models of, a universe in which all reduced information and causal power lay at one length-scale – one could label it the 'sub-atomic'. An urgent question is whether the universe we live in belongs to that class, or to its complement. The alternative possibility belongs to universes in which there is no privileged length-scale, but within which the 'fundamental' emerges at all length-scales. Put in mathematical terms, reductionism insists that irreducible degrees of freedom are spread broadly in real-space co-ordinates, but concentrated within an extremely narrow band of reciprocal space. An emergent cosmos possesses fundamental, constitutive and independent degrees of freedom as widely dispersed in reciprocal, as in real space. The latter is both far less arbitrary than the former, and immensely more consonant with the way that physics works. 'The completeness of physics' has, in philosophical parlance, become unfortunately synonymous with 'the completeness of sub-atomic physics', which is by no means the same thing. If physics tells us anything, it is that no attempt to comprehend the world at a single length-scale works, but that one can identify 'fundamental' components at emergent length-scales (Ellis 2012). A deeper understanding of mind will need to recognize that it is by no means the 'next-stop' in a causal hierarchy from quantum fields upwards.

The questions of significance, agency, ethics and reciprocity lead, it seems to me, beyond *Scientia* to *Sophia*. In the reconciliation, renewal and reinvention of the future human with the cosmos we will need more wisdom than knowledge. In so far as wisdom has implicit and explicit roots, the search for her leads us inexorably into theological questions. Should a theological *Sophia* guide and inform a scientific *Scientia*? Dirk Evers warns against it, and advocates the tending of the garden fence between science and theology. He is right to warn against mixing epistemologies, but the problem with Steven J. Gould's non-overlapping *magisteria* is that that science has always required a nurturing medium from which to draw motivation, imagination, courage and purpose. We might cite Grosseteste once more, as spokesman for the medieval renaissance of natural philosophy:

> Since sense perception, the weakest of all human powers, apprehending only corruptible individual things, survives, imagination stands, memory stands, and finally understanding, which is the noblest of human powers capable of apprehending the incorruptible, universal, first essences, stands! *(Grosseteste 1992, 167)*

The purpose of natural philosophy was directed towards a restoration of lost knowledge of the natural world though the fall. Surprisingly for those who have disconnected their knowledge of modern science from its early modern roots, far from embodying some secular awakening from an infertile scholastic past, this is also the narrative of early modern science, as advanced by Francis Bacon. There is a direct line of succession within this underground stream of irrigation of science by theology to the vision of Fedorov and his Cosmists.

Markus Mühling proposes a framing of an engaged relationship consonant with the experience of scientists and that goes beyond the familiar categories of Ian Barbour. In particular, he points out that a 'dialogue' presupposes a contest, even if a weak one, over a shared subject. I am receptive to his preference of the metaphor of a walk together across hilly terrain in which sometimes one partner, sometimes the other, leads. His immersive, experienced theo-philosophy of science, that through its echoes of Heidiggerian *Dasein*, manages to resolve the question begged by all the positive Barbour categories – namely the daily experience of the sciences in interpreting observations *etsi deus non daretur*. In other words, this is the challenge, for a theistic scientist, of the efficacy of methodological naturalism.

To extend the metaphor, however, we ought to think about the maps and guides the posse of science, theology and philosophy takes with them on the trail, even on hikes into unexplored territory. Bob White reminds us that the Bible, for Christians, has held a special place in the knapsack for this very reason. Theologian N. T. Wright describes how the 'authority' of the Biblical inheritance might be taken to act on believers today in a similar way as the Shakespearian corpus would be authoritative over a company of actors, immersed in his plots and prose, now attempting to improvise the lost final act of a newly-discovered play from the Bard's pen (Wright 1995). White likewise points out that there are important motifs and ideas in the Psalms and Prophets that speak to a divine care for non-human creation on the one hand, and the destructive consequences within physical nature of human disobedience on the other. It is not necessarily so very large a leap from his application of climate change to, say, considerations of 'climate change' in a deliberate sense, on other planets.

Elsewhere, I have suggested that the tributaries of the natural sciences themselves lie in the Biblical Wisdom tradition (see McLeish 2014). For if we look there for material that treats the human relationship with the created material world, immediately the texts pour forth like a river. The first thing to notice is the frequency with which the creation story is told and retold: recall where scriptural narrative refers back to God's act of creating the world: Proverbs 8, Psalm 19, Psalm 33, Psalm 104, Isaiah 40, Isaiah 45, Jeremiah 10, Hosea 2, and in the New Testament John 1 provides a salient example, are just a few of the places where different language, a rich variety of metaphor or fresh pictures are used to remind God's people that it was their Lord who laid the foundations of the Earth, separated the land and the sea, spread out the heavens. To take two examples a little further: the delightful and playful creation account in Proverbs 8 amounts to the story of wisdom's birth – here she (*Sophia*) is a little girl at the feet of the Creator, playing with the rivers and mountains.

> The Lord brought me forth as the first of his works, before his deeds of old;
> I was appointed from eternity, from the beginning, before the world began.
> When there were no oceans, I was given birth,
>   when there were no springs abounding with water,
>   before the mountains were settled in place, before the hills, I was given birth,
>   before he made the earth or its fields, or any of the dust of the world.
> I was there when he set the heavens in place, when he marked out the horizon on the face of the deep,
>   when he established the clouds above and fixed securely the fountains of the deep,
>   when he gave the sea its boundary so that the waters would not overstep his command,
>   and when he marked out the foundations of the earth.
> Then I was the craftsman at his side.
> I was filled with delight day after day, rejoicing in his presence,
>   rejoicing in his whole world and delighting in mankind.

The Lord is certainly in the background, but centre stage is the fleeting, dancing character of wisdom who weaves herself into the components of the physical world, responding to their coming into being with joy. Although very old, this complex passage already contains several themes that reappear in or lie closely below the surface of a long skein of creation literature in the Bible. At its heart it embeds a sort of 'formula' for describing the world's creation that concentrates on establishing boundaries: '*marked out the horizon ... established the clouds above ... gave the sea its boundary ... marked out the foundations of the earth*'. The vital aspect of creation in this tradition is not so much the naked existence of matter, but its *order*: the sky above, the sea separated from the land and the depths of the earth below our feet. The 'taming' of the sea is especially important – a dangerous and alien medium is kept at bay almost as one would a tethered beast.

More is true: creation stories, wherever they occur in scripture, tend to form bridges from a position of hopelessness and lost-ness to a renewed hope. So the great recapitulation of creation in Isaiah 40 leads directly to the announcement of the One who will come to redeem Israel. Psalm 33 takes a (brief) journey through the creation of the cosmos to take the psalmist from despair to hope. Some of these accounts are very short, which in turn tells us just how developed the two Genesis creation stories are (in Chapters 1 and 2 – by different authors, using different language and metaphor), but brevity does not imply insignificance. Human relationship with the physical creation is also a growing theme in these recurrent motifs – the celebration of the wisdom of the farmer who knows which seeds to plant at what season is the focus of Isaiah 28. People are not called to sit back and contemplate physical creation, but to engage with it fruitfully

Perhaps the most profound of all the wisdom scriptures, in its description of the human relation with the natural world, is the enigmatic *Book of Job*. Contemporary neo-Kantian philosopher Susan Neiman has recently even urged that *Job* be held alongside Plato as a foundational text for western thought (Neimann 2016). I have never tired of losing myself in this wonderful book since first I fell captive to what must surely be the greatest poem of natural wisdom in all ancient literature – the so-called Lord's Answer of chapters 38–42. Here, for the first time since the book's prologue, the Lord finally appears to Job in answer to his repeated demands for vindication and for God's admission that his suffering is unjust. But rather than tackling Job's complaints head-on, Yahweh takes the man on a journey through all of creation, and at every waypoint asks him a *question*:

> Have you journeyed to the springs of the sea?...
> Where is the way to the abode of light?...
> ...From whose womb comes the ice?...
> ...Do you know the laws of the heavens?
> And can you apply them to the earth?

Scientists to whom I have recommended the reading of these chapters have always come back astonished – for here are the foundation questions of the sciences we now call 'meterology', 'oceanography', 'cosmology', 'astronomy' and zoology'. More than that, as all working scientists know, the vital step in all successful science is not the finding of the correct answer (in spite of the years of science schooling that would have us believe so) but the formulation of the creative question. Einstein and many other creative scientists have noted this. Heisenberg, for example, wrote, 'In the course of coming into contact with the empirical method, physicists have gradually learned how to pose a question properly. Now, proper questioning often means that one is more than half way towards solving the problem.'

Strangely, 'The Lord's Answer' has received some tough criticism in the scholarly literature. On the one hand it is charged with irrelevance – Job is concerned with the moral issue of the suffering of the righteous, not the provenance of the snow or the lightning. On the other, God is accused of the petulant put-down – of suggesting by his list of unanswerable questions that Job is ignorant and should cease his complaining. Neither objection holds on close reading, however. For one thing, the entire *Book of Job* is replete with nature imagery. All the animals, plants and phenomena referred to in 'The Lord's Answer' have already been invoked in the three cycles of discourses between Job and his friends over the first thirty-seven chapters. Job's complaint is in fact a double one: he accuses God of allowing chaos to reign in the natural world just as much as he does in the moral world:

> What he destroys will not be built, whom he imprisons will not be freed.
> He holds back the waters, there is drought; he lets them loose, they overwhelm the earth. (Ch12)

Here we meet an example of the nature imagery that threads throughout the book – chaotic floods introduced here reappear in the questioning Lord's Answer. It is there because Job accuses the Lord of creating physical chaos in the world, as he does moral chaos in Job's suffering. As for the reason for God's appearance, far from diminishing Job, he is invited to 'stand up' and debate on Yahweh's level, as in a courtroom. The vital context for the long questioning poem is the earlier 'intermission' to the cycle of discourses in chapter 28 often called the 'Hymn to Wisdom'. Mysteriously beginning under the earth, down a mine, it follows the miners as they 'dangle and sway' on their ropes, looking up at the earth from beneath. The author wonders that, of all the creatures, only human eyes are able to see the inner structures of the earth in this way. Not even the falcon, 'with her sharp eyes', can see right into the structure of the rocks that are open to the eyes of the miners. Then the depths of earth and sea are questioned on where wisdom can be found – without avail. The Hymn ends with identifying wisdom as a divine way of seeing:

> But God understands the way to it; it is he who knows its place.
> For he looked to the ends of the earth, and beheld everything under the heavens,
> So as to assign a weight to the wind, and determine the waters by measure,

The miner's eyes peering into the deep structure of the world from the glimmer of a lamp becomes a faithful metaphor for science itself – that part of culture that develops our gift of seeing beneath the surface of phenomena in the light of observation, imagination and reason. But there is more, for this close of the chapter indicates that it is in just this ability that we are made in the image of God as regards Wisdom, for this 'deep seeing' into the world is what Wisdom is, and what the Creator does. This is the deeply immersed, contemplative, sense in which *etsi deus non daretur* is both utterly the appropriate *experience*, while totally mistaken as *interpretation* of the experience of natural philosophy. (I cannot bear to call it 'science' any more.) We do not see God in creation because we look into it, in love, co-creation and care *with Him.*

Take the ancient invitation to Job, and thereby to all who follow him, to engage in a deep and questioning way with the natural world, including the utterly trans-human and 'other' of Behemoth and Leviathan, constitutes a theological guidebook, and a narrative framework, for a challenging future in a vast universe.

References

Davies, S., P. Macnaghten and M. Kearnes (eds), *Reconfiguring Responsibility: Deepening Debate on Nanotechnology* (Durham: Durham University, 2009).

Dupuy, J.-P., 'The Narratology of Lay Ethics', *Nanoethics* 4 (2010): 153–70.

Ellis, G., 'Top-down Causation and Emergence: Some Comments on Mechanisms', *Interface Focus* 2 (2012): 126–40.

Evans, J. S. B. T., and K. E. Stanovich, 'Dual-Process Theories of Higher Cognition: Advancing the Debate', *Perspectives on Psychological Science* 8 (2013): 223–41.

Grosseteste, Robert, *Commentary on the Posterior Analytics*, quoted in R. W. Southern *Robert Grosseteste; The Growth of an English Mind in Medieval Europe* (Oxford: Clarendon Press, 1992), 167.

Hugh of St Victor, *De tribus diebus 4* (PL 176, 814B).

McLeish, Tom, *Faith and Wisdom in Science* (Oxford: Oxford University Press, 2014).

McLeish, Tom, 'The Search for Affirming Narratives for the Future Governance of Technology: Reflections from a Science-Theology Perspective on GMFuturos', in *Governing Agricultural Sustainability*, ed. P. Macnaghten and S. Carro-Ripalda (Oxon: Routledge, 2015).

Neimann, Susan, The Rationality of the World: A Philosophical Reading of the Book of Job, *ABC net* (2016). http://www.abc.net.au/religion/articles/2016/10/19/4559097.htm

Poincaré, Henri, 'Mathematical Creation', in *The Foundations of Science*, trans. G. B. Halsted (Lancaster, PA: The Science Press, 1915).

Stanovich, K. E., and R. F. West, 'Discrepancies between Normative and Descriptive Models of Decision Making and the Understanding Acceptance Principle', *Cognitive Psychology* 38 (1999): 349–85.

Wilkinson, David, *Science, Religion and the Search for Extraterrestrial Intelligence* (Oxford: Oxford University Press, 2013).

Wright, N. T., 'The New Testament and the People of God', in *Perspectives on Psychological Science* 8 (London: SPCK, 1995): 223–41.

Young, George M., *The Russian Cosmists: The Esoteric Futurism of Nikolai Fedorov and His Followers* (New York: Oxford University Press, 2012).

Zola, Emile, 'The Experimental Novel', in *The Naturalist Novel*, ed. Maxwell Geismar (Ste. Anne de Bellevue: Harvest House Ltd, 1964).

INDEX